ELECTRIC MOTORS
Second Edition

By the same author
Workshop Practice Series 24
Electric Motors in the Home Workshop
ISBN 978 185486 133 7

ELECTRIC MOTORS
Second Edition
Jim Cox

SPECIAL INTEREST MODEL BOOKS

Special Interest Model Books Ltd.
P.O.Box 327
Poole
Dorset
BH15 2RG

First published by Argus Books Ltd. 1988

Reprinted 1990, 1991, 1992, 1993, 1994, 1995, 1996, 1998, 2000, 2002, 2003

Second edition published by S.I.Model Books 2006

Reprinted 2008

ISBN-13 978-185486-246-4

www.specialinterestmodelbooks.co.uk

Contents

Introduction

Small workshops can use a wide variety of electric motors ranging from fractional horsepower motors for machine drive to tiny micromotors in small mechanisms. Successful installation and operation of these motors needs a reasonable understanding of the operating characteristics and limitations of the wide range of motor types that are now available.

Much of the generally published information on motors is very basic in nature and is of little help in using motors in non-standard or unusual situations. While more information exists in specialist journals and textbooks it is too deeply and obscurely buried to be useful to the non-specialist user.

This book is intended to fill this gap by setting out in simple form the essential characteristics and operating limitations of the principal motor and generator types. The approach is essentially practical in nature with few calculations needing anything more than simple arithmetic.

Basic operation and installation information is given for the first time user. In addition to this more detailed information is included to enable the advanced user to get the best out of motors in unusual and demanding applications.

The main part of the book covers the operating characteristics of motor types commonly encountered in domestic and workshop machinery. Advice is given on identification of the ratings of unknown motor types and on how to use domestic, automotive and industrial motor types in small workshop applications.

A major section deals with a number of methods of operating industrial three phase machines from domestic single phase supplies and compares the advantages and weaknesses of different systems.

In addition to this, sections are included on less common motor and generator types such as servo motors and stepper motors. Readily available in the surplus market, the unusual types can be extremely useful in special applications.

Control and installation problems are covered and this includes operation both from European 240/415 volt 50Hz and North American 115/230 volt 60Hz supplies. The control section includes data on motor starting systems, electronic speed control and motor braking systems.

FOREWORD

Safety

Electricity is in every home and is so useful and normally so safe that we tend to take it for granted. So long as the high voltages are protected by properly insulated cables and terminations the chance of electric shock is remote and this is the normal situation in home use.

However, if live conductors are exposed or the connected equipment is faulty the normal 240 or 115 volt domestic mains supply is quite capable of giving the user a very unpleasant, or in extreme cases fatal, electric shock. It is essential therefore to use safe working practices when wiring up, installing or testing mains voltage electric motors and equipment.

The following guidelines should always be observed:−

1. Switch off *AND* unplug from the mains before touching any conductor that might be live in normal operation. If it is a permanently wired circuit which cannot be unplugged then switch off and remove the main fuse supplying that circuit. Keep the fuse in your pocket to ensure that no-one else can put it back before you are ready. It is *NOT* sufficient to just switch off without also disconnecting the circuit. As any

insurance company will tell you, faulty switches are not particularly rare events and you don't want to find out the hard way. In the case of permanently wired circuits it is also good practice to use the blade of a well insulated screwdriver to short together the input conductors before you allow your fingers near them − just to make sure that you have switched off the right circuit and pulled the right fuse.

2. The strength of an electric shock is determined by the amount of current that flows and not directly by the voltage. In the human body much of the resistance to current flow is in the thin layer of dry skin which covers the rest of our fat and muscle which are soaked in body fluids. Anything which penetrates or wets (including perspiration) this layer of skin enormously increases our susceptibility to electric shock, so try to keep your hands dry and cover any cuts or abrasions.

The key point to bear in mind is to avoid the possibility of any electric shock where the current path is through the chest as this can upset the heart muscle. The common danger paths are hand to hand, hand to foot and hand to head.

Damp concrete floors are quite good conductors so if you are working on this sort of surface wear rubber-soled shoes to prevent a hand to foot path.

Hand to head paths are relatively uncommon – most of us instinctively avoid touching things with our head. The main hazard is the head touching the metal parts of an earthed/grounded reading lamp or bench light. The cure here is either to use a lamp with no exposed metal parts or to use a modern double insulated reading lamp. These have two core instead of three core flex leads and the doubly insulated external metal parts are not connected to ground.

Hand to hand is the most difficult possibility to exclude and this underlines the importance of carefully observing the precautions outlined in 1. above.

Professional electricians, on some occasions, have to work on live circuits which cannot be disconnected. They are used to the meticulous care that this demands and one technique which is used is never to touch the work with more than one hand at a time, the other hand being kept safe not touching any metallic object. Readers of this book are emphatically *NOT* advised to work on live circuits; however, the principle of avoiding two hand contact where practicable is a useful safeguard in addition to normal precautions.

3. Always connect to earth the motor

A selection of electric motors covering many of the types described in this book, from small D.C. to industrial units of various designs

frame or equipment casing - even on temporary test rigs. This is good practice on any equipment but doubly important on items of dubious origin where the fault may well be an intermittent failure of the insulation between windings and case!

4. Before applying power to a motor or similar device do make sure it's properly anchored to something solid. When a motor starts, the reaction to the starting torque can cause it to leap off the bench with obvious electrical and mechanical hazards.

5. Be sure you understand the correct connections of the motor/circuit that you are working on. In subsequent chapters of this book advice is given on testing, operation and installation of motors and allied equipment in hopefully a clear and understandable form. However, it is not possible to anticipate all eventualities —so if you are in any doubt don't take risks - consult a qualified electrician.

6. Some capacitors (see section 1.8 and 3.4.3) used in conjunction with motors may retain a charge after the motor has been switched off. This charge may persist for many hours unless the capacitors are fitted with discharge resistors. Always discharge capacitors by shorting together the terminations with the blade of an insulated screwdriver before touching the associated wiring. Small or low voltage capacitors are usually innocuous but anything over about 4mF at 240v or 16mF at 115v should be treated with caution.

7. The above comments refer particularly to equipment operating from 240 volt domestic mains supply. 115 volt supplies are a little more forgiving as the hazardous currents are roughly halved but still need to be treated with respect.

Below 50 volts shocks are rarely hazardous unless exceptionally low resistance contact is made or the individual is particularly susceptible to shocks of any kind.

At 6 to 24 volts, which is typical of automobile and model activity, the main hazard is thermal as the power source may be capable of delivering large short circuit currents which can raise connecting wires to red heat in an embarrassingly short time. If working on an automobile battery do not wear a wristwatch with a metal wristband. It is natural to rest the wrists on the battery when making connections and if a metal wristband should happen to bridge terminations a very nasty burn can result in seconds.

8. Industrial motors are rated for maximum winding temperatures in the range 100°C/210°F to 165C°/330°F. The casing of the motor will be considerably cooler than the windings and these temperatures are only reached at full load and maximum ambient temperature. Nevertheless, be careful as there are plenty of occasions when outside parts of a motor will be hot enough to raise a nasty burn if grasped incautiously.

9. Dependent on your location there may be special rules governing installation and usage of electrical or electronic items. Be sure that your pet project does not infringe these.

CHAPTER 1

A Few Basics

1.1 General Comments

Perhaps the best thing about this chapter is that it isn't essential reading. If you skip it and move straight to the more interesting bits practically everything will make sense. However, if you want a bit of background on some of the terms and explanations that come later, then half an hour or so spent on this chapter may save a little head-scratching.

In the rest of the book a fair amount of trouble is taken to go beyond bald statements of fact and to explain how things happen and why they happen. It is not possible to take that approach in this chapter. Here we are trying to cover some of the essential elements of a two-year course in basic electrics in a matter of a few pages. In the space available the best that can be done is to set out some of the main controlling concepts in as simple a way as possible. More detailed information can be found in the basic physics and electrical engineering sections of your local library.

1.2 Ohms Law

This defines the relation between voltage, current, and resistance in an electric circuit (figure 1-1).

(1) $E = I \times R$

(2) $I = E/R$

(3) $R = E/I$

E = Voltage in volts (E for Electro-Motive Force)

I = Current in amps (I for Intensity of current)

R = Resistance in ohms

Power in an electrical circuit is measured in Watts or Kilowatts. One Kilowatt is one thousand watts. With powers stated in watts the relationships are:—

(4) $P = E \times I$

(5) $P = I^2 R$

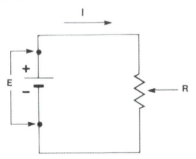

FIGURE 1.1   ELECTRIC CIRCUIT

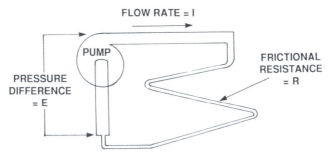

FIGURE 1.2   PUMP/PIPE EQUIVALENT

(5) is only a restatement of (4) with E replaced by I × R. It is a very frequently used relationship because it states directly the power lost in a resistor carrying a particular current. In motor circuits there are often several different types of loss occurring at the same time. The term I^2R loss is often used to describe the power loss arising from current flowing through the resistance of a winding to distinguish it from other types of loss.

These are the basic relationships for D.C. (Direct Current) circuits, i.e. a circuit fed from a battery or other fixed polarity source. A rough analogue of what is happening is a pump driving water round a long wiggly pipe (figure 1-2).

The pressure difference between the pump inlet and outlet is equivalent to voltage.

The friction between the flowing water and the pipe is equivalent to resistance.

The rate of flow is equivalent to current.

The pressure difference at the pump times the flow rate is equivalent to power.

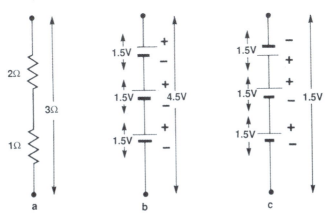

FIGURE 1.3   SERIES CONNECTION

11

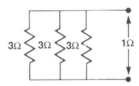

FIGURE 1.4   PARALLEL CONNECTION

1.3 Series connection

If a second resistor is connected "in series" with the first resistor (figure 1-3a) then the total circuit resistance is the sum of the values of the two individual resistors.

In the same way, if a second battery is connected in series with the first, the total voltage is the arithmetic sum of the two voltages. Arithmetic sum means that you must take polarity into account. If Positive on one battery is connected to Negative on the second battery the voltage of the second battery adds to the first. If Positive is connected to Positive the voltage of the second battery subtracts from the first.

1.4 Parallel connection

If two or more resistors are connected in parallel (figure 1-4) then their conductances add. Conductance is the reciprocal of resistance i.e. 1 divided by the value of the resistor. Scientific calculators have a reciprocal button usually marked 1/x. Most (not all) of the simpler calculators can be persuaded to behave in the same way by first entering the number, press "divide" twice, press "equals" twice. Once the conductances have been added, the reciprocal of this value is the resistance of the parallel combination of resistors. This operation is expressed as:–

$$(6) \quad R = \frac{1}{1/R_1 + 1/R_2 + 1/R_3 \text{ etc.}}$$

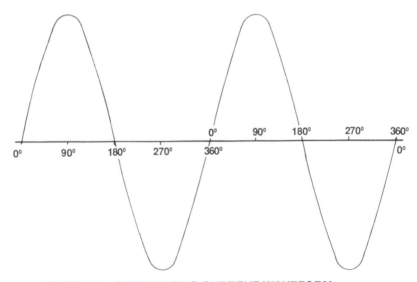

FIGURE 1.5   ALTERNATING CURRENT WAVEFORM

For just two resistors in parallel the following expression does the same job with fewer key presses:–

(7) $$R = \frac{R_1 \times R_2}{R_1 + R_2}$$

1.5 A.C. Circuits

If the polarity of the battery is reversed many times per second then the direction of the current it supplies to the load alternates at the same rate, i.e. Alternating Current. In practice alternating current is usually generated by rotating machines called alternators. In these, the polarity does not suddenly reverse but follows a smooth curve from a maximum positive value to a maximum negative value (figure 1-5). The shaft of the simplest type of alternator rotates through three hundred and sixty degrees to deliver one complete positive and negative cycle. The output waveform is called a sine wave because the amplitude of each point on the waveform is directly related to the sine of the angle that the alternator has reached in its three hundred and sixty degree rotation.

It is very important to understand this relation between time and angle. The waveform that is plotted in figure 1-5 is a plot of voltage on the vertical axis against time on the horizontal axis. However, because we know that the parent alternator rotates through three hundred and sixty degrees in one complete cycle it is equally valid to calibrate the horizontal axis in degrees. This is a great convenience because if we think in terms of electrical degrees this always defines the same point on the waveform irrespective of the alternator speed/supply frequency.

Domestic power supplies are almost always A.C. because it is easier to generate and distribute than D.C. The most common standards are 220 or 240V 50Hz in Europe and 110V or 115V 60Hz in North America. (One Hertz (Hz) is one cycle per second).

Apart from specialised applications the voltage of an A.C. supply is always specified as the R.M.S. value (root mean square) which is 70.7% of the maximum value of the sine waveform. This 70.7% value is chosen because in most circuits it has the same heating power as a D.C. voltage of the same level. This means that Ohms law applies without alteration.

1.6 Inductance

If a voltage E is applied to a coil of wire of resistance R the current will eventually be equal to E/R and the current flow will generate a magnetic field in the space surrounding the coil. A coil which can generate a magnetic field has the property of inductance.

The magnetic field is generated by the current in the coil but, as the magnetic field grows, it opposes the rate of growth of current so that the magnetic field and the current do not reach their final value immediately. The magnetic field cannot prevent the growth of current – only slow it down. The current reaches about two thirds of its final value in L/R seconds (L is the inductance of the coil in Henries, R the resistance in Ohms).

With D.C. applied, the current always reaches the final value E/R and the inductance of the coil does not affect the final current at all.

However, if A.C. is applied, long before the current can reach its final value, the polarity of the applied voltage has reversed and is trying to change the current in the opposite direction. The

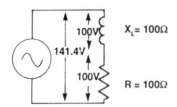

FIGURE 1.6 A.C. CIRCUIT

inductance behaves as a sort of A.C. resistance which appears in series with the normal D.C. resistance of the coil. The value of this A.C. resistance is given by:–

$$(8) \quad X_L = 2\pi FL$$

X_L = Inductive reactance in ohms
F = Frequency in Hz (i.e. cycles per second)
L = Inductance in Henries

These inductive ohms (the proper title is reactance) behave in a similar but not identical way to D.C. ohms (i.e. resistance) in an A.C. circuit. The reason for this is that reactance ohms are lossless and do not dissipate power. The

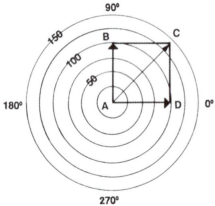

FIGURE 1.7 VECTOR ADDITION

resistance to current flow is caused by energy being stored in the rising magnetic field in one part of the cycle and then being returned without loss to the circuit by the collapsing magnetic field in the next part of the cycle. This interchange results in current in an inductance lagging behind the applied voltage. In a pure inductance (i.e. a coil with zero resistance) the peak value of the current appears 90 degrees later in the cycle than the peak value of the applied voltage. This is called a 90° lag relationship. Because of this, reactive ohms do not add directly to resistive ohms and the impedance of a series circuit containing both resistance and reactance is given by:–

$$(9) \quad Z = \sqrt{X^2 + R^2}$$

Z = Impedance in ohms
X = Reactance in ohms
R = Resistance in ohms

The impedance Z is the effective value of X and R and determines the current that will flow in response to an applied voltage.

This indirect method of series addition applies only to mixed resistance and reactance. Series-connected inductors behave in exactly the same way as series-connected resistors and add directly to each other.

1.7 Vectors

Figure 1-6 shows the interesting paradox that in an A.C. circuit the sum of the voltages across the individual elements can exceed the applied voltage. For the values shown, (9) tells us that the total circuit impedance is 141.4 ohms so that 1 Amp flows through X and R. Since X and R are both 100 ohms 100V will appear across each of these components and the total will well exceed the supply voltage.

Figure 1-7 shows how this can be represented. The length of each arrow is equal to the size of the voltage or current being represented. The direction represents the phase angle. These arrows are called vectors.

The vector of length 100 in the direction of 0° represents the voltage across R. The vector of length 100 in the direction 90° represents the voltage across L. Because L and R are in series the same current flows through each so, by definition, the current through each has the same phase angle. The voltage across R is in phase with the current so the voltage vector points to 0°. We know that the current through a pure inductance lags the current by 90°. Since the currents are in phase the voltage across L leads the voltage across R so this vector points to −90°.

The sum of these two vectors is found by completing the parallelogram ABCD. The sum is then the diagonal vector length AC of 141.4V. The angle of this vector also indicates the phase angle of the sum voltage – in this case −45°.

This is a very useful way of simultaneously visualising the voltage and the phase relations between different A.C. voltages or currents. We shall use it again when we look at three phase supplies and induction motors.

1.8 Capacitance

Two metal plates separated by an insulating material have the property of capacitance. Insulating material used in this way is usually called a dielectric.

If the metal plates are connected to a voltage source, current will flow into the plates and produce a voltage stress in the dielectric. When this voltage stress equals the input voltage no further current will flow. The capacitor (sometimes called condenser) is now charged up to the input voltage and if disconnected the plates will remain charged up to that voltage.

A perfect capacitor would remain charged for ever but, in practice, imperfections in the dielectric allow the charge to slowly leak away. Commonly used capacitors hold their charge for periods varying from a few minutes to many hours. This leakage is so small that its effect can be ignored when using capacitors with motors.

Figure 1-8 shows a voltage V applied to a capacitor C through a resistor R. If C is initially uncharged, the initial capacitor voltage will be zero and all the voltage will appear across R. The current flowing through R starts to charge C and, as the voltage rises, the voltage across the resistor falls by a corresponding amount. Steady state is reached with no current flowing through the resistor and all the voltage across the capacitor.

The capacitor cannot prevent the voltage across a-b reaching supply voltage – it can only slow it down. The voltage reaches about two thirds of its final value in C×R seconds (C is the capacitance in *FARADS*, R is the

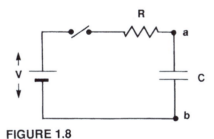

FIGURE 1.8

CAPACITOR/RESISTOR NETWORK

resistance in ohms). The Farad is the fundamental unit of capacitance and is inconveniently large for everyday use. Capacitors used with motors will normally be stamped with the value in microfarads (μF). One μF is one millionth of a Farad. If μF and ohms are used for the CR product then the time interval is in microseconds (μS).

If an A.C. source is used, long before the voltage on the capacitor can reach its final value, the polarity of the supply voltage reverses and tries to charge up the capacitor in the opposite direction. This means that the current is maintained and the capacitor behaves as an A.C. resistance (i.e. reactance). The value of this reactance is given by:-

$$(10) \quad X_c = \frac{1,000,000}{2\pi FC}$$

X_c = capacitative reactance in ohms
F = Frequency in Hz (i.e. cycles per second)
C = capacitance in μF

Note – The 1,000,000 corrects for the use of the capacitor value in μF instead of Farads

Capacitors in series and parallel connection behave differently to inductors or resistors. Parallel connection *adds* the value of all the parallel connected capacitors.

If capacitors are connected in series then their admittances add (admittance is the reciprocal of capacitance). The rules for dealing with this are exactly the same as for parallel connected resistors (see section 1.4).

As can be seen from the following table, capacitance is almost the inverse of inductance.

	R	L	C
Impedance at D.C.	R	Zero	Infinity
Impedance at Frequency F	R	$2\pi FL$	$\dfrac{1}{2\pi FC}$
Series connected	$R = R_1 + R_2$	$L = L_1 + L_2$	$C = \dfrac{C_1 \times C_2}{C_1 + C_2}$
Parallel connected	$R = \dfrac{R_1 \times R_2}{R_1 + R_2}$	$L = \dfrac{L_1 \times L_2}{L_1 + L_2}$	$C = C_1 + C_2$
Power loss at current I	$I^2 R$	Zero	Zero
Current lead or lag on applied voltage	$0°$	$+90°$	$-90°$

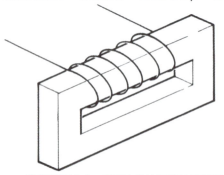

FIGURE 1.9   COIL ON IRON YOKE

1.9 Permeability and Magnetic circuits
Both the inductance of a coil and the total amount of magnetic flux generated by a given current depend on the magnetic permeability of the space surrounding the coil. Air and most other substances have a permeability factor of 1. A few substances, mainly iron, nickel and cobalt, are what is known as ferromagnetic. These have a very much higher permeability – in the range 1,000 to 100,000.

16

If all the air around the coil is replaced by ferromagnetic material both the inductance of the coil and the total magnetic flux is increased by the permeability factor. The electrical grades of iron used in motors have permeabilities of several thousand. Because of this high permeability, if a coil is wound round an iron yoke (figure 1-9), almost all the magnetic flux passes through the iron – the iron acts as a magnetic conductor. Unfortunately iron can only carry a limited amount of flux before it starts to saturate. The permeability then becomes lower and lower for each increase in current until eventually no significant increase in flux can be obtained however large the current. The iron is then said to be saturated.

The maximum torque that a motor can produce is proportional to the total magnetic flux that crosses the air gap between rotor and stator. Because of the excessive amount of current needed to operate the iron near saturation flux density most of the iron in an electric motor has to work well below saturation level. This limits the amount of torque that can be produced by a given size of motor.

Magnetic circuits behave in a rather similar way to electric circuits:–

Voltage is replaced by ampere-turns (i.e. the number of turns x the current flowing through the coil).

Current is replaced by total magnetic flux.

Resistance is replaced by a new term – reluctance. Reluctance is determined by the permeability, and the area to length ratio of the iron circuit i.e.

$$(11)\ S = \frac{l}{a \times \mu}$$

S = Reluctance
a = cross sectional area
l = length
μ = Permeability

(11) is included to show the meaning of the term "reluctance" and how it is composed. It does not specify the units of measurement because these depend on other measurement conventions which are beyond the scope of this book.

Permanent magnets are roughly equivalent to an electro-magnet with a fixed number of ampere-turns per centimetre length "frozen" into the iron. Modern permanent magnet materials can exceed the performance of an electromagnet of similar size and weight. Because of this motors with permanent magnet fields are a little smaller and lighter than their wound field counterpart.

1.10 Generator and Motor action
If a wire moves at right angles to a magnetic field a voltage will appear at the two ends of the wire. The size of the voltage will depend on the total amount of magnetic flux cut per second but it doesn't matter how this is distributed along the length of the wire. For maximum voltage generation the wire should move at high speed through a strong magnetic field. The polarity of the output voltage is determined by the direction of motion relative to the field. If the polarity of the field is reversed then the output voltage reverses.

In small machines a single wire rarely develops sufficient voltage at an acceptable speed so a number of wires are connected in series. It is usually possible to arrange these connections in the form of a continuous coil. Figure

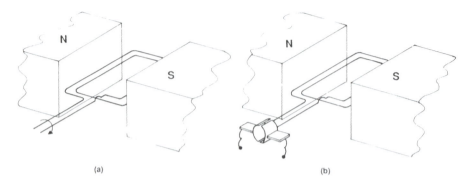

(a)   (b)

FIGURE 1.10   A.C. AND D.C. GENERATORS

1-10a shows a machine using a continuous coil rotating in a transverse magnetic field. With this arrangement, when rotation starts, the voltages generated in all four go-and-return wires add in series. Initially a large voltage is generated as the wires are moving at right angles to the magnetic field but, as the coil rotates, the angle decreases until at 90° the wires are moving parallel to the field. At this point the flux cut per second has dropped to zero so the output voltage is also zero. As the rotation continues the angle becomes more favourable with the voltage reaching a second maximum at 180°, this time with the polarity reversed.

With a uniform magnetic field, the flux cutting rate changes with the sine of the angle of rotation so that the output voltage follows the sine curve shown in figure 1-5.

If the wires of figure 1-10a are connected to slip rings, brushes bearing on these slip rings can connect the alternating voltage to an external circuit. If D.C. is required the wires can be connected to a two segment commutator as shown in figure 1-10b. This automatically reverses the connections to the two brushes each time the output voltage reverses.

The output voltage is now always of the same polarity but of very variable level (figure 1-11a). This variation (usually called "ripple") can be reduced to almost any desired level by using more coils and more commutator segments. Figure 1-11b shows the large reduction that occurs when the commutator switches between three coils at 120° intervals.

So far the discussion has been entirely in terms of voltage. As long as no output current flows, the wires can move freely in the magnetic field. No output power is delivered and no power is required to turn the coil.

The situation is quite different once current flows through the wires. If a current is passed through a wire running at right angles to a magnetic field, a sideways force will be generated proportional to the product of the strength of the current and the total magnetic flux crossing the wire. If the current is caused by a load resistor connected to the machines described above, the direction of the force will oppose the rotation. In a perfect

18

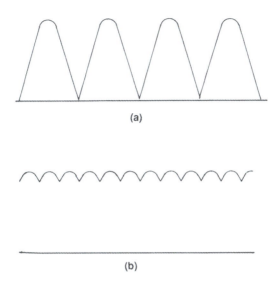

(a)

(b)

FIGURE 1.11   GENERATOR FITTED WITH COMMUTATOR

machine the mechanical power needed to overcome this force will equal the electrical power delivered to the load.

If the mechanical drive is disconnected and the rotating element held still with a spring balance, a current passed through the coil will exert a torque on the shaft. The torque will fall to zero if the rotor is allowed to rotate to the 90° position but, if the machine is provided with multiple coils and commutator segments to keep changing the coil in use, the torque can be maintained throughout the full 360° rotation. This enables it to operate as a motor.

In a perfect motor, if the rotor is rotating it will be generating a voltage which will oppose the applied voltage (usually called the "back E.M.F."). If no mechanical load is applied the two voltages will be equal and no current will flow. If a mechanical load is applied the load will try to slow the motor down and just sufficient current will flow to generate the required torque. This results in the inverse of the generator case — mechanical output power now equals electrical input power.

1.11 Reluctance machines

All the discussion so far has been on the effects arising from a current carried by a moving conductor immersed in a magnetic field. This is the basic element of most motors and generators.

However, it is also possible to build Variable Reluctance machines which rely on projections sticking out from a soft iron rotor which vary the reluctance of the path taken by the main magnetic field as the rotor turns within the stator.

In a generator the output voltage is generated because the resultant variation in field strength through the output winding is equivalent to the winding encountering the same change in flux as a result of moving through a fixed field.

In a motor the self-generated voltage which opposes the input voltage is generated by the same means. The output torque is generated by magnetic attraction between the stator and the tips of the rotor projections. This magnetic attraction occurs because, if parts of a magnetic circuit are free to move, they will always take up the position of minimum total reluctance because this results in maximum total flux.

1.12 Transformers and Auto-transformers

These are devices for changing (transforming) the power supply voltage to a value which suits the user equipment. They will only work on A.C. supplies and a small amount of power is lost in the process – typically less than 10%. Apart from this loss, output power equals input power but at a different voltage.

If a number of coils are wound round a leg of an iron yoke their magnetic fields interact (figure 1-12). If an A.C. voltage is applied to one of the coils it will induce a voltage in the other coils. The winding to which the voltage is applied is called the "primary", the remaining coils are called "secondaries".

This is called a transformer and the induced voltage in each secondary is the primary voltage multiplied by the turns ratio, primary to secondary. A transformer can have any ratio "step up" or "step down". If 100V is applied to a 1,000 turn primary, 300 turn and 5,000 turn secondaries would have output voltages of 30V and 500V.

An auto-transformer operates in exactly the same way as a transformer but uses a single tapped winding which acts as both primary and secondary. A step down auto-transformer uses the whole winding as the primary with the output taken between one end and the tap. A step up auto-transformer reverses the connections with the whole winding used as the secondary.

An auto-transformer uses its copper more efficiently than a transformer and, because of this, is typically half the size and weight of a transformer of similar power rating.

Transformers are wound for particular voltages. Most transformers will stand 10 to 15% more than their nominal voltage; beyond this the iron starts to saturate (see section 1.9) and the transformer will overheat.

The voltage ratio will only equal the turns ratio when there is no load on the secondary. As soon as a current is taken from the secondary, a voltage drop occurs due to the resistance of both the primary and the secondary windings. Transformers are very efficient and the full load output voltage will be typically 90 to 97% of the theoretical value.

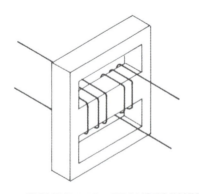

FIGURE 1.12   TRANSFORMER

1.13 Relays and Contactors

Relays and contactors are simply electrically operated switches. Relay is the generic term: the term contactor is reserved for relays that switch relatively high powers.

The relay/contactor moves its switching contacts when rated voltage or rated current is applied to the operating coil. It is current that is important because it is current that generates the magnetic field that causes the contacts to move. Rated voltage is simply the voltage that is necessary to produce that current in the resistance of the operating coil.

Relays may have only a single on/off contact or up to about half a dozen changeover contacts (a changeover contact is a moving contact which touches one contact when the relay is not energised and moves to a second contact when it is energised). Contactors often have two or more heavy duty "make" contacts to switch on the main power circuits plus one or more sets of light duty contacts for control circuit switching. Typical relays suitable for motor control are shown in figure 1-13.

1.14 Diodes and Rectifiers

These are devices that will only pass current in one direction – the electrical equivalent of a non-return valve in a water pipe. Both terms are used interchangeably to describe the same device. The term diode is mostly used to describe single elements operating at low power levels. The term rectifier is used to describe diodes or assemblies of more than one diode that are intended to be used in power circuits to convert A.C. into D.C.

Figure 1-14 shows an assortment of the type of diodes/rectifiers likely to be encountered in association with motors

Fig. 1.13 *Relays for motor control*

Fig. 1.14 *Diodes and rectifiers*

and control gear. The actual rectifier elements are very small chunks of silicon buried in a protective case which is usually black plastic but occasionally glass or metal.

The two main types are single elements which have two axial wires and bridge rectifiers which are four tag devices containing an assembly of four diodes. The circuit symbols are shown in figure 1-15; current flows from a positive input in the direction of the black arrow to the cross bar of the symbol. When used in a rectifier circuit the cross bar end will be a source of positive voltage and this end is marked with the symbol + or by a distinguishing band of colour.

The main use of these devices in connection with motors is to convert single phase A.C. power to D.C. If a single diode is placed in series with the supply (figure 1-16), alternate half cycles are suppressed and the output is unidirectional (i.e. D.C.) but with a very large amount of ripple. The average value of this D.C. output is a little less than half the A.C. input voltage (0.45 x Vin R.M.S.).

Figure 1-17 shows that with a bridge rectifier both half cycles of the input A.C. can be steered to the output + and − terminals. This doubles the average value of the output to 0.9 Vin and reduces the amount of ripple.

In electronic applications capacitors and inductors can be used to smooth out this ripple component to give a pure D.C. output. However, motors are pretty tolerant of ripple and smoothing is not normally necessary.

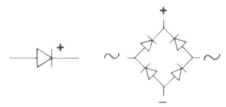

FIGURE 1.15   DIODE AND BRIDGE RECTIFIER SYMBOLS

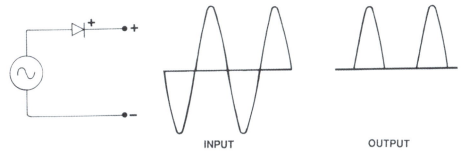

INPUT   OUTPUT

FIGURE 1.16   HALF WAVE RECTIFIER

1.15 Single Phase and Three Phase Power

So far, only single phase A.C. power has been considered i.e. the sort of power produced by the single coil of the alternator shown in figure 1-10a. If two more coils are added spaced out at 120° intervals the alternator will have three sine wave outputs. Because of the angular spacing of the coils each successive output will reach its maximum 120 electrical degrees later than the earlier output. These three outputs are called phases and the 120° difference in timing is called a 120° phase lag.

Three phase power has a number of key advantages over single phase.

a) For a given power rating three phase motors and generators are 20-30% smaller than their single phase counterparts. This is because it is possible to interleave and evenly distribute the three windings around a rotor and stator. A single phase winding has to be located in two short sectors so that the winding distribution is uneven and cannot make full use of the available space. In addition to this the three phase magnetic flux pattern makes more efficient use of the iron in the stator.

b) By using the three phases in a motor winding a true rotating magnetic field can be generated. This means that three phase motors are self-starting. Single phase motors are not self-

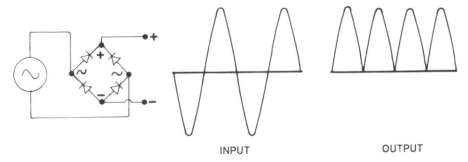

INPUT   OUTPUT

FIGURE 1.17   FULL WAVE RECTIFIER

23

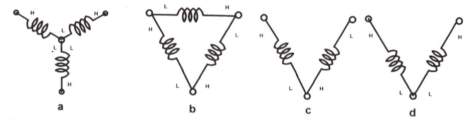

FIGURE 1.18 FOUR COMMON THREE PHASE CONFIGURATIONS

starting and need auxiliary starting circuits.

c) When transmitting power over long distances three phase circuits can use less copper than single phase. If separate go-and-return conductors are used for each of the three phases then, for the same transmitted power, three phase would require the same total weight of conductors as single phase. However, a four wire system is normally used with all three phases sharing a common return conductor. If an equal load is placed on each phase (i.e. a "balanced load"), the three currents 120 degrees out of phase cancel and no current flows in the return conductor. In practice, allowance must be made for some load unbalance but, even so, substantial savings are possible in the size of the common return conductor.

This method of connection is usually called the Y or Star form of connection as shown in figure 1.18 a.

It is not the only possible three phase configuration. Figure 1.18 shows four common connections.

a) This is the already described star form - Because the three generating voltages are mutually 120 degrees apart the voltages do not add directly but add to give the line to line vector sum of $\sqrt{3}$ x generator voltage

b) The same generators with the same mutual phase difference but reconnected in the Delta format of b also produce a balanced (i.e. equal line to line voltages) three phase system. This time the line to line voltage is equal to the generator voltage.

Note the direction of the generator connections. In Star, all the generator lows are common but in Delta each generator high connects to the next generator low. This reversal is exactly equivalent to a 180degree additional phase shift which is why, in the Delta arrangement, generating voltages 120 degrees apart results in voltage vectors 180 - 120 = 60 degrees apart which form an equilateral triangle.

Although three correctly phased generator voltages are needed to form a fully symmetrical three phase Delta connection, any one of these can be removed and a balanced set of three phase voltages remains. This is the Open Vee system shown in c and d.

c) Shows the Open Vee that is formed by removing one leg from a normally connected 120-degree apart three phase generating system. The single common line is a generator low connected to the adjacent generator high.

d) This is exactly the same system but this time the generator winding on the right is reversed so that it is the generator lows

24

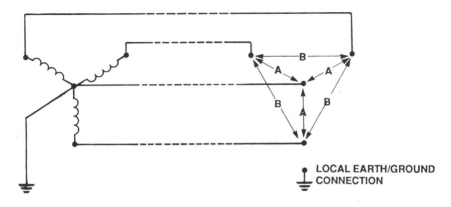

LOCAL EARTH/GROUND
CONNECTION

A/A/A/ 220/240V PHASE TO NEUTRAL DOMESTIC USER SUPPLIES

B/B/B/ + NEUTRAL 380/415V LINE TO LINE 3 PHASE INDUSTRIAL USER SUPPLY

FIGURE 1.19 EUROPEAN POWER DISTRIBUTION

that are common. This then requires only 60 degrees phase difference between the two voltages. This latter arrangement is very useful when it is necessary to produce synthetic three phase supplies from a single phase source see section 3.3.1.

It is very important to remember that this removal of one generating leg of the supply still permits correct supply of power to all three line connections. It is NOT the same as loss of a line connection from a three phase source - e.g. a blown fuse. If one line of a three phase connection is lost all that is left is two lines and two lines can only supply single phase power.

In all the above figures the end of each winding is identified by an L or an H to indicate which way round it is connected. As drawn all four systems have the same direction of phase rotation – any of them, connected in the same way, to a three phase motor would cause it to rotate in the same direction. If any pair of wires

connecting the three phase source to the motor is reversed this is exactly equivalent to connecting to a mirror image of the source. This means that the direction of phase rotation and the consequent motor rotation is reversed.

The Star, Delta and open Vee connections of a, b and c are the commonly encountered arrangements based on voltages initially generated 120 degrees apart. The open Vee of d with 60 degree separation is a special case normally only found in single to three phase converters.

In the U.K. and most of Europe electric power is distributed at user level in Star as four wire three phase 380/220 or 415/ 240 volt power. Figure 1-19 makes this clear, 220 or 240 volts is the voltage between the central neutral wire and any of the three live phase wires, 380 or 415 volts is the voltage between any pair of the three live phase wires. The neutral wire is connected to earth at the Electricity Generating Board distribution point.

Domestic users are supplied with one neutral wire and one phase wire which results in a 220 or 240 volt single phase supply. Any of the three phase wires delivers 220/240 volts so the electricity company connects roughly equal numbers of houses to each of the three phases to achieve a balanced load on the supply lines.

Industrial users are supplied with all three phase wires and the common neutral wire. Although this is exactly the same 220/240 volt phase to neutral supply it is usual to refer to it as a 380/415 volt supply because this is the voltage when measured between any two of the live phase lines (this is known as the LINE voltage).

It is relatively easy to upgrade domestic users to a three phase supply because the necessary extra lines are readily available from the local distribution transformer.

In all the above the 220/240 and 380/415 volt ratings were quoted because these were commonly used nominal voltages in different parts of Europe and also the usual motor nameplate ratings. Common Market standardisation has now resulted in a change to a rather wide tolerance 230 volt nominal single phase / 398 volt three phase supply.

North American domestic power points are 110 or 115V 60Hz single phase. However, this is one side of a three wire 110 - 0 - 110V or 115 - 0 - 115V supply connection, which delivers 220 or 230V between the two outer wires. High power loads such as domestic cookers can be wired directly to the full 220/230V. Although three wires are used, this is a dual voltage single phase supply, not a three phase system.

There is less standardisation in the methods used to distribute three phase power to light industrial users. The most popular arrangement is a delta or open Vee connected 230V line to line supply. To make available single phase outlets, one of the legs is configured as a 115-0-115V supply with the centre point grounded. For higher power users the delta connection is retained but the line to line voltage increased to 460V

Motors intended for use on these supplies are provided with two sets of windings, which can be interconnected to enable 230V, or 460V operation.

1.16 Mechanical matters

Generators convert mechanical power into electrical power. Motors make the opposite conversion - electrical power into mechanical power.

Mechanical power can be measured in Horse-Power or as its power equivalent — mechanical watts or mechanical kilowatts. If the motor mechanical output is dissipated in a friction load then 1kW of mechanical power produces exactly the same amount of heat as a 1kW electric fire. The Horse-Power is a slightly smaller unit - 1 H.P. is equal to 746W; 1kW is equal to 1.34 H.P.

It is important to avoid confusion between motor rated input power and mechanical output power. Industrial motor nameplates state input power as motor voltage and full load current.

Mechanical output power is stated in kW or H.P. This is pretty unambiguous.

Domestic machines have suffered from the advertisers' desire to quote the biggest and most impressive number. If you are offered a vacuum cleaner fitted with a "900 Watt motor" this is almost certainly the full load input power. The actual mechanical output power depends on the efficiency of the motor and may be anywhere between about 400W and 700W.

Motor output power is proportional to speed x torque. Speed is normally stated in revolutions per minute (R.P.M.). Torque can be stated in a variety of units. The following units are used in this book:-

British units - pound inches (lb in) or ounce inches (oz in)

Metric units - kilogram centimetres (kg cm) or gram centimetres (g cm)

In each case the unit is a force acting at a radius of either 1 inch or 1 centimetre. These are directly measurable practical units. Torque can also be stated in Newton metres.

The relationship between torque, speed and power is given in (12) and (13)

$$(12) \quad H.P. = \frac{R.P.M. \times lb\ in}{63000}$$

$$(13) \quad kW = \frac{R.P.M. \times kg\ cm}{97400}$$

$$(14) \quad 1\ kW = 1.34\ H.P.$$

CHAPTER 2

Induction Motors

2.1 Introduction

Induction motors are the most commonly used type of motor in industry. They enjoy this popularity because of their simplicity, reliability and, in most cases, high efficiency.

They all rely on the use of sets of windings in a fixed stator to generate a rotating magnetic field to drive a rotor which provides the mechanical output. Because the stator can only generate a rotating field if supplied with alternating current these motors cannot operate from direct current supplies and their maximum output shaft speed is limited by the supply frequency to 3,000 R.P.M. at 50Hz in Europe or 3,600 R.P.M. at 60Hz in North America.

Fig. 2.1 *Induction motor rotor and stator*

2.2 Construction and Cooling

The key parts of an induction motor are the laminated iron stator which carries the windings which generate the rotating magnetic field and a laminated iron rotor containing a pattern of copper or aluminium conductors which constrain the rotor to follow the rotating field. A typical example of each is shown in figure 2-1. These elements are fitted in a light alloy or steel housing which also carries the rotor bearings and the electrical terminations.

The basic motor may be mounted in a number of different ways. The most popular are foot mounting, resilient mounting or flange mounting shown in figures 2-2, 2-3 and 2-4. The foot mounting and resilient mounting types are normally used when belt drive is used to couple the motor shaft to the load. The resilient mounting type takes up a little more space than foot mounting but it is used when it is important to minimise the noise and vibration transmitted by the motor through the frame. This is often necessary with single phase motors as these inherently generate a vibration component at twice the supply frequency. Three phase motors do not suffer from this problem and are normally solidly mounted. If the motor is directly coupled to the load or to a gear train, flange mounting is used as this makes it much easier to achieve the necessary precise positioning of the output shaft.

The motors in figure 2-2 and 2-3 are open ventilated types. These have air

Top, Fig. 2.2 *foot-mounted motor*
Centre, Fig. 2.3 *resilient mounted motor*
Bottom, Fig. 2.4 *flange mounted motor*

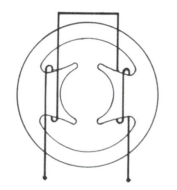

FIGURE 2.5   SINGLE PHASE STATOR WINDING

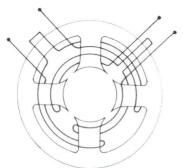

FIGURE 2.6   TWO PHASE STATOR WINDING
EACH COIL SURROUNDS TWO POLE PIECES

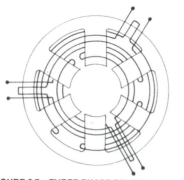

FIGURE 2.7   THREE PHASE STATOR WINDING
EACH COIL SURROUNDS THREE POLE PIECES

vents at each end of the housing and an internal fan or projections on the rotor draw air through the motor to cool the windings. In industrial environments it is undesirable to allow possibly polluted air to pass through the motor and the totally enclosed fan ventilated construction shown in figure 2-4 is preferred. Here the motor windings are totally enclosed and cooling is provided by an external fan blowing air over the motor casing. To provide sufficient cooling it is usually necessary to fin the outer casing of the motor.

The permissible operating temperature of a motor is determined by the type of insulation used in the windings. Industrial motors are normally rated for operation in a maximum ambient temperature of 40°C/105°F. At full load, motors with Class B insulation might have an internal temperature rise of 80°C/145°F so the peak winding temperature can reach 120°C/250°F. Only about half this temperature rise reaches the outside casing of the motor but the surface temperature can still be over 80°C/175°F – much too hot to touch! Class F insulation motors run even hotter, 100°C/210°F rise with a surface temperature near boiling point.

Motors running on part load at normal room temperature are perhaps more commonly encountered and may feel no more than comfortably warm. However, as the above figures show, a motor can be running well within its rating and still be hot enough to cause a nasty burn.

2.3 Induction Motor Characteristics
Induction motors are provided with stators wound for operations from single phase, two phase or three phase supplies. In principle larger numbers of

phases could be used but there is no practical advantage in exceeding three.

In order to achieve the most efficient arrangement of both the copper and iron circuit the windings are distributed in multiple slots in the laminated iron stator. The stator is built up from a stack of castellated rings (laminations) punched from thin sheets of electrical grade iron. Each lamination is insulated from the next by a thin layer of varnish. This prevents the stator iron acting as a shorted turn within the windings. If a solid stator were used it would behave as a partly short circuited turn coupled to the windings and the eddy currents induced would give rise to serious losses. The layers of varnish between the laminations in a laminated stator break up the current path and prevent this happening.

It is common for each of the stator windings to be distributed between a number of stator slots. However, as far as the rotor is concerned they behave as single windings, one per phase, each generating a field transverse to the rotor as shown diagrammatically in figures 2-5, 2-6 and 2-7 for one, two and three phase stator windings.

Figures 2-8, 2-9 and 2-10 show how the current in each winding changes with time through one complete cycle of the supply frequency for each of the three cases together with the direction of the resultant magnetic field at the different points in the supply cycle. The three winding arrangements give broadly similar performance but since the three phase case is the closest approach to the ideal it is easier to look at this first.

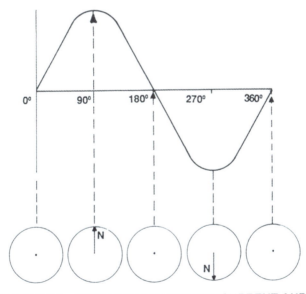

FIGURE 2.8   SINGLE PHASE STATOR CURRENT AND
MAGNETIC FIELD VECTORS

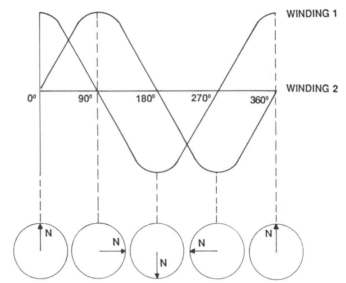

FIGURE 2. 9   TWO PHASE STATOR CURRENTS AND
MAGNETIC FIELD VECTORS

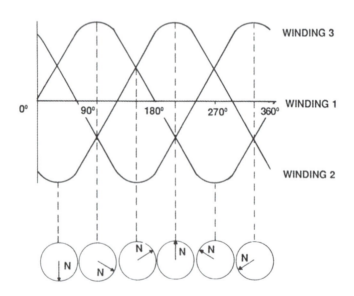

FIGURE 2.10   THREE PHASE STATOR CURRENTS AND
MAGNETIC FIELD VECTORS

The three phase case generates a true constant amplitude rotating field i.e. the sum of three fields generated by the three stator windings is constant and the direction rotates through 360° in one cycle of the supply frequency.

The rotor takes the form of an iron cylinder threaded with a symmetrical arrangement of short circuited conductors (usually called a squirrel cage rotor because of its appearance) and the rotating magnetic field generated by the stator induces large currents in the short circuited conductors. The interaction between the magnetic field generated by these currents and the main rotating field generates a torque on the output shaft.

The strength of the current induced in the rotor is determined by the rate at which the field is changing within the rotor and is large when the rotor is stationary, falling to zero when the rotor speed equals the rotating field speed. This means that the normal squirrel cage rotor machine can never reach synchronous speed – the rotor speed must always be sufficiently lower than the rotating field speed (the "slip speed" or "slip frequency") to induce sufficient current in the rotor conductors.

Most induction motors are fitted with low resistance rotors (i.e. large section conductors) as this permits large induced currents to flow in the rotor conductors when the rotor speed is only slightly less than synchronous speed. At this low "slip" frequency the rotor inductance has little effect and almost all the circulating current produces useful torque at the output shaft. However, if, for any reason, the rotor speed is reduced by a large amount the "slip" frequency increases and at this higher frequency the rotor inductance becomes a major part of the rotor impedance and this both limits the induced current in the rotor conductors and moves the effective magnetic field pattern so that only part of the induced current produces useful output torque.

This results in the motor characteristic shown in figure 2-11. At synchronous speed there is no relative movement between the rotor conductors and the rotating field generated by the stator so that the output torque is zero. As the rotor speed

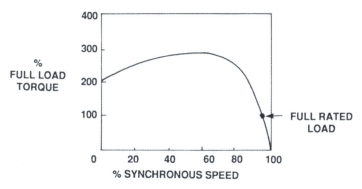

FIGURE 2.11   THREE PHASE MOTOR TORQUE/SPEED CURVE

drops (increasing the slip frequency and rotor current) the torque generated by the induced rotor currents rapidly increases. With about 1% slip enough torque is generated to overcome the motor no-load mechanical losses giving a no load speed of close to 99% of synchronous. The motor rated full load torque is typically reached at 95% of synchronous giving an almost constant speed characteristic – only 4% change from no load to full load.

If the load on the motor output shaft is increased further, as the speed drops, the output torque increases to a peak of two or three times the full load torque at somewhere between half and two thirds synchronous speed. At lower speeds, although the circulating current in the rotor is still increasing, the effect of the rotor inductance becomes dominant because of the high slip frequency and the available torque reduces, dropping to about twice rated full load torque at zero speed. This is the "starting torque" (sometimes called "locked rotor torque") and is the maximum load torque that the motor can overcome and accelerate towards its rated full load speed.

Although the motor can deliver somewhat more torque at its peak torque speed, the speed range between this and zero speed is potentially unstable. If any attempt is made to load the motor in excess of this peak torque capability the motor speed will drop and reduce the available torque. This is a cumulative effect, each drop in speed further reducing the available torque until the motor abruptly stalls. The motor can only operate stably in this part of the speed torque curve if it is driving a load whose torque requirements reduce rapidly as speed drops. Fan loads are typical of this class (the torque load

that a fan places on its driving motor increases as the square of the speed) and this is one of the reasons that makes it possible to control the speed of an induction motor driving a fan over a wide range by varying motor input power.

It must be emphasised that motor operation below its rated full load speed is only permissible during starting or temporary overloads. Continuous operation in this region will result in overheating unless arrangements are made to reduce the motor input power.

2.4 Multi Pole Motors

All the above examples are based on the basic two pole motor i.e. each phase winding induces a single pair of North and South poles in the rotor and this single pair of poles rotates through one complete revolution in one cycle of the supply frequency. These are known as two pole motors and have an output shaft speed about 5% less than the synchronous speed which is 3000 R.P.M. on 50 Hz supplies or 3600 R.P.M. on 60 Hz supplies.

If lower speeds are needed, multi pole stator windings can be used. These are categorised by the number of North and South poles produced round the circumference of the stator. Because, at the full load speed, the rotor will move through just less than two poles in one cycle of the supply frequency the following output shaft speeds result:-

	Synchronous Speed RPM		Full Load Speed RPM	
Supply	50Hz	60Hz	50Hz	60Hz
2 Pole	3,000	3,600	2,850	3,450
4 Pole	1,500	1,800	1,425	1,725
6 Pole	1,000	1,200	940	1,140
8 Pole	750	900	700	850

The two and four pole winding arrangements make the most efficient use of the iron and copper circuits and are by far the most popular. As the number of poles increases beyond four, for a given horsepower rating, the motor becomes larger, more expensive and less efficient.

Six and eight pole machines are often used for large low speed fans to avoid the complication of a belt or gear speed reducer between motor and fan. However, in the more common case of the motor coupled to the load by a speed reducing belt drive, a two or four pole motor is the most efficient arrangement.

There is no fundamental limit to the number of poles - motors with thirty or more poles are made but they are only used where low shaft speed is a primary aim and efficiency and size secondary considerations.

2.5 Single Phase Motors

Because motors for home use must operate from the single phase domestic supply we must forgo the luxury of the almost pure rotating field that a three phase winding can generate and put up with the limitations imposed by the rotating component of a single phase oscillating field. There are several different ways of visualising the rotating component of an oscillating field but they all lead to the same conclusion – it will work fine once the motor is up to running speed but the starting torque will be zero.

One way of looking at it is to note that if we superimpose a clockwise rotating field on an anti-clockwise field the result is a single phase oscillating field. This checks with the observed facts – if a single phase motor is spun up to operating speed in either direction it will

continue to run equally well in that same direction, clockwise or anti-clockwise.

While this is a perfectly valid way of visualising the rotating component of a single phase oscillating field and much loved by the text books, it stretches the imagination somewhat and it is probably easier to consider it as one component of a two phase field. In a similar manner to the three phase system described above a properly distributed two phase field can produce a true constant amplitude rotating field. If we look at figure 2-9 we can see that at the 0 and 180° points in the cycle winding 1 provides all the magnetic flux, winding 2 is making no contribution because at these times the current in winding 2 is zero. 90° later, in each case, the position is reversed with all the flux coming from winding 2 and none from 1. If we now move to the single phase case by omitting winding 2 it is easy to see that the correct flux conditions for a rotating field occur twice per cycle of the supply frequency and that zero rotating field also occurs 90° later twice per cycle. This results in the torque developed in the rotor varying between zero and the maximum value twice per cycle of the supply frequency. This is exactly what happens in a single phase motor although it is not normally evident because the torque fluctuations are almost completely smoothed out by the mechanical inertia of the rotor. However, a single phase motor will produce noticeably more noise and vibration than a similar three phase motor because of this effect.

The torque versus speed characteristic of a single phase motor is shown in figure 2-12. Between no load and full load the torque curve is very similar to the equivalent three phase

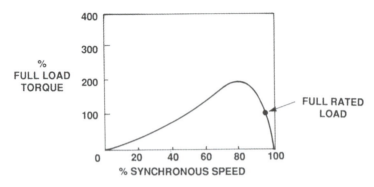

FIGURE 2.12   SINGLE PHASE MOTOR TORQUE/SPEED CURVE

motor and in the same way the torque rises to an overload peak somewhat below full load speed. However, at lower speeds the torque falls away rapidly until at zero speed there is no torque at all. This means that all single phase motors must have some mechanism for accelerating up to a speed far enough up the torque/speed curve to be capable of generating full load torque. In most cases this means about two thirds full load speed. Many methods are possible but all the popular ones rely on auxiliary windings which, when switched into circuit during starting, convert the motor into an approximation to a two phase motor. The starting windings are switched out of circuit when the motor reaches operating speed. True two phase operation is difficult because only a single phase is available from the supply. Fortunately all that is necessary is to produce a reasonable fraction of full load torque throughout each cycle of the supply frequency. This can be achieved by fitting a starting winding wound in stator slots 90° displaced from the main winding and supplied with current some tens of electrical degrees leading or

lagging the main winding current. It doesn't matter whether the starting winding current leads or lags the main winding because this only changes the direction of rotation.

2.6 Split Phase Motors
The simplest starting method is known as "Split Phase" starting. It uses a main winding filling most of the stator slots and an auxiliary starting winding of fewer turns partly filling some of the slots at 90° to the main winding. Because the main winding of the motor is wound with many turns surrounded by the iron circuit and fills many slots it has a high self-inductance and a relatively low resistance. This ratio of inductance to resistance (L/R) controls the phase angle between the supply voltage and the winding current and results in the phase angle of the current in the winding lagging the supply voltage phase angle. The auxiliary starting winding is wound with fewer turns. The inductance is so low that the starting winding current is almost in phase with the supply voltage and this provides the necessary difference in phase between the

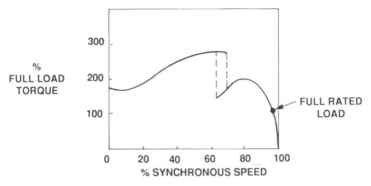

FIGURE 2.13   SINGLE PHASE MOTOR TORQUE/SPEED
CURVE WITH STARTING WINDING

currents in the two windings. The difference in phase is well short of the ideal 90° and the inductance of the starting winding is so low that very large currents flow during starting – seven to ten times normal full load current is typical. A large fraction of this power is dissipated in the starting winding and to avoid catastrophic overheating it is necessary to disconnect the starting winding as soon as the motor has run up to speed – at most a second or so after starting. This is normally carried out by a centrifugal switch mounted inside the casing and operated by a spring loaded pair of weights mounted on the rotor shaft. The switch contacts are closed when the rotor is stationary and remain closed until the rotor reaches approximately 75% full load speed when the centrifugal force on the weights is large enough to overcome the spring force and open the switch contacts to disconnect the starting winding.

This results in the speed/torque curve shown in figure 2-13. The dotted discontinuities are the switching points of the centrifugal switch. The right hand discontinuity is where the centrifugal switch opens during normal motor starting. The left hand discontinuity is where the switch recloses if the motor is severely overloaded – this must never be allowed to happen in normal use as the power then dissipated in the windings may be as high as fifty times normal and the windings will overheat in seconds.

The starting torque delivered is typically 1½ to 2 times full load torque which is ample for most small workshop machines e.g. drill presses, lathes, milling machines and grinders. However, unless the motor is run a long way below its maximum temperature and power ratings it is necessary to avoid frequent stops and starts which would overheat the starting winding. If frequent stops and starts are unavoidable it is better to change to a capacitor start motor (described later) or to leave the motor running continuously and stop and start the load with a clutch.

The split phase motor is most popular in the power range below ½ H.P. (370w). At power levels above this the very high starting currents cause difficulties in the control and protection gear.

37

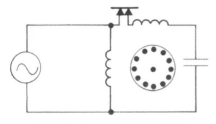

FIGURE 2.14   CAPACITOR
START MOTOR

2.7 Capacitor Start Motors

These differ from split phase motors in that the starting winding has many more turns (frequently more than the main winding) and is fed via a series capacitor, see figure 2-14. The result of this slight additional complication is much better starting characteristics. The series capacitor causes the current in the starting winding to lead the supply voltage phase angle and, with correct choice of winding and capacitor, the ideal 90 degree phase difference can be closely approached. The capacitor start motor has both a lower starting current and provides more starting torque than the equivalent split phase motor. A typical torque/speed curve is shown in figure 2-15. Starting torque is now two to three times full load torque with starting current values of four to six times full load current.

To achieve this performance rather large values of capacitance are needed – typically $50\mu F$ or more per horsepower at 240v 50Hz and about four times larger for 115v 60Hz. The only type of capacitor which can provide this sort of capacity and voltage rating within an acceptable size and cost is an electrolytic capacitor. This is a variety of capacitor in which the charge is stored in an extremely thin insulating anodic layer electrolytically formed on pure aluminium foil. One electrode of the capacitor is the aluminium foil, the other electrode is a conducting liquid in contact with the anodic film. This provides the necessary large capacitance in a small volume but unfortunately suffers from small but significant series and shunt losses which lead to serious internal heating when carrying large alternating currents. This is not too important when used as a starting capacitor (most

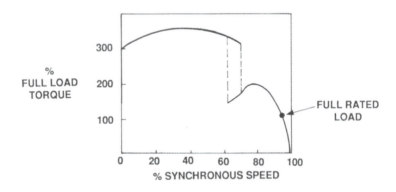

FIGURE 2.15   CAPACITOR START MOTOR
TORQUE/SPEED CURVE

motors, if severely maltreated, would burn out their starting windings long before the capacitor got hot enough to matter) but rules out its use as a permanently in-circuit capacitor in a capacitor run motor (see next section).

The capacitor start motor is the ideal general purpose single phase workshop motor. It has lots of starting torque, will tolerate frequent stop/start operation and is only marginally more expensive than the basic split phase machine. It can be recognised readily by the characteristic bulge usually at 2 o'clock on the motor casing which houses the cylindrical starting capacitor.

2.8 Capacitor Run Motors

Sometimes called Permanent Split-phase Capacitor Motors, these are the logical extension of the capacitor start motor but with the capacitor remaining in circuit all the time. This eliminates the centrifugal starting switch but generates a fresh set of problems which limits the use of these motors to fairly specialised applications.

The first problem is the change of impedance of the windings as the motor accelerates from stationary rotor to full speed. When the rotor is stationary the rotor conductors behave as a shorted turn closely coupled to the stator windings and this results in a low winding impedance. As the rotor speed increases this effect reduces until at full speed the winding impedance is three or more times higher. The optimum value of capacitor changes in similar ratio so that the capacitor can be chosen for best starting performance or best run performance but not both.

The second complication is the capacitor itself. The electrolytic capacitors used with capacitor start motors are not suitable for continuous operation and it is necessary to use capacitors specially designed for continuous A.C. operation. These capacitors usually use polypropylene or oil impregnated paper as the dielectric and are very much larger and more expensive than the equivalent electrolytic capacitor.

Because of this the motors are mainly optimised for the run condition with a high impedance (i.e. more turns) capacitor phase to reduce the amount of capacitance needed. Because at full speed the motor is running as almost a true two phase machine they are quieter and smoother than most single phase motors.

On the debit side these motors have very poor starting torque — rarely as much as full load torque and sometimes as low as one fifth full load torque. Even to achieve this starting performance it is often necessary to use a high resistance rotor design which results in a higher full load "slip" frequency so that the full load shaft speed is only some 90% of synchronous instead of the 95% achieved with normal low resistance rotors.

These motors are mainly used for fan drives as these do not need a high starting torque, or for very small motors where there is insufficient space to house a centrifugal starting switch.

2.9 Capacitor Start/Capacitor Run Motors

These use a large starting capacitor to give good starting torque and, as soon as the motor is up to speed, switch to a smaller value to give optimum "run" conditions. This combines the good starting characteristics of a capacitor start motor with the smooth running of a capacitor run motor. This type of motor is relatively uncommon and is mainly used in the larger sizes of single phase motor for applications where the smoother running and improved power factor are a real advantage.

2.10 Shaded Pole Motors

All the motors described so far use a multi-turn starting winding. The shaded pole motor is different in that the starting winding is in circuit all the time and takes the form of one or two copper loops encircling part of each stator pole. This "shades" part of the pole from the main stator field and the current induced in the loop causes the field generated by this shaded part to lag the main field. The phase shift is less than the ideal 90° and the strength of the shaded field considerably less than the main field. Because of this the starting torque is poor, typically only half full load torque. Considerable power is dissipated in the

shading loops which are in circuit all the time and this results in low efficiency. Full load efficiency is rarely as high as 20% and the part load efficiency of a small motor may be as low as 2 or 3%. This also results in poor speed regulation and lower rated full load speed. The following figures are typical:—

	R.P.M.	
	50Hz	60Hz
2 Pole	2,550	3,100
4 Pole	1,275	1,550

In spite of this lack-lustre performance, the shaded pole motor is made in very large quantities because of its simplicity, very low cost and suitability for low power ratings. Power outputs range from 1 to 50W (0.001 to 0.07 H.P.) and at these low power levels the low efficiency is rarely a problem. Nevertheless, because of their high losses, shaded pole motors always run very hot, even at no load.

Fig. 2.16 *An example of a shaded pole motor*

40

Fig. 2.17 *Shaded pole motors in two applications*

The larger 2 pole and 4 pole machines use simple circular stator laminations with slots in projecting pole pieces for the shading rings (in most cases the "ring" is in fact rectangular), see figure 2-16. When assembled with its bearing housings it does not look very different from its larger split phase cousins. However, in the smaller sizes a radically different form of construction is used to reduce manufacturing cost.

Figure 2-17 shows a motor of this type. The pole piece and shading ring arrangement is very similar to the previous type but the twin magnetic return paths of the circular laminations are replaced by a single asymmetric larger return path. This enables the twin specially shaped coils needed with circular construction to be replaced by a single bobbin wound coil. To maintain the necessary close tolerance on the bore and circularity of the rotor tunnel it is common practice to allow a narrow

neck of lamination material to mechanically bridge the two pole pieces. In operation, because of the small cross section, this part of the magnetic circuit is saturated and only a small part of the total flux flows through this unwanted alternative path.

This type of motor is often used to drive fans in freezers, fan heaters or pumps in washing machines. The odd ex-washing machine pump can be pressed into service as a coolant pump but, apart from this and light duty fans, these motors are not particularly useful in the small workshop as most of the low power applications are better suited to small commutator motors.

2.11 Reversing the Direction of Shaft Rotation

The shaft rotation direction of three phase motors can be easily reversed by interchanging any two input wires. This reverses the direction of field rotation and is effective both from rest and from when the motor is already running in the opposite direction. However, caution should be exercised when reversing a motor from full speed (sometimes called "plugging"). The motor reversal current may considerably exceed the normal starting current and the total extra heating in the motor is at least equal to two normal starts in quick succession.

A capacitor run motor behaves rather like a three phase motor with poor starting torque. It can be reversed by interchanging the connections to the capacitor phase, or by interchanging the connections to the main winding – it doesn't matter which.

Split phase and capacitor start motors can be reversed by interchanging the connections to the starting winding, or by interchanging the connections to the main winding. However, this is only effective when starting from rest – once the motor is up to speed the starting winding is completely disconnected by the centrifugal switch and reversing it makes no difference to the motor which will continue to run in the same direction.

In shaded pole motors the direction of rotation is determined by the position of the shading rings – the rotor will always turn towards the shaded edge of the pole pieces. The only way to reverse a shaded pole motor is to take it to pieces and then reassemble it with the rotor turned end for end in the rotor tunnel. This is not normally a difficult operation, usually no more than removing and replacing a pair of bolts.

2.12 Dual Voltage

Machines intended for international use are frequently fitted with dual voltage motors capable of operating from 220/240V 50Hz or 110/115V 60Hz supplies.

These motors are provided with two sets of main windings which can be connected in series for 220/240v operation or, by rearranging links on the motor terminals, connected in parallel for 110/115V operation.

In principle two sets of starting windings could be provided for similar series and parallel connection. However, in practice, it is sufficient to fit a single set of 110/115V starting windings. For 110/115V operation these are connected directly across the supply via the starting switch and capacitor (Figure 2-18b). For 220/240V operation they are reconnected to between one side of the supply and the junction point of the two series connected main windings (Figure 2-18a). In this connection the main windings act as an auto-transformer reducing the 220/240V input voltage to the 110/120V required by the starting winding.

The full load speeds for 50 and 60Hz operation will be slightly different as indicated in the table in section 2.4 but this difference is rarely enough to matter.

In the 110/115V connection the reversing arrangements are exactly the same as for a single voltage motor i.e. a double pole changeover switch is needed to reverse the connections to the starting winding. In the case of the 220/240V connection of the dual voltage motor the same method is normally used. However a simpler arrangement is possible because the centre tap at the junction of the main windings winding makes it possible to reverse the direction of current in the starting winding by a single pole changeover switch – Figure 2-18c shows the arrangement.

The terminal cover plate will normally show the terminal links for the two operating voltages but if this is missing it is reasonably easy to sort out the connections with a few resistance measurements and a little trial and error. With all links removed the resistance of the three windings can be measured. Two windings will have almost exactly the same resistance – these are the main windings. The resistance of the third winding will be significantly different –

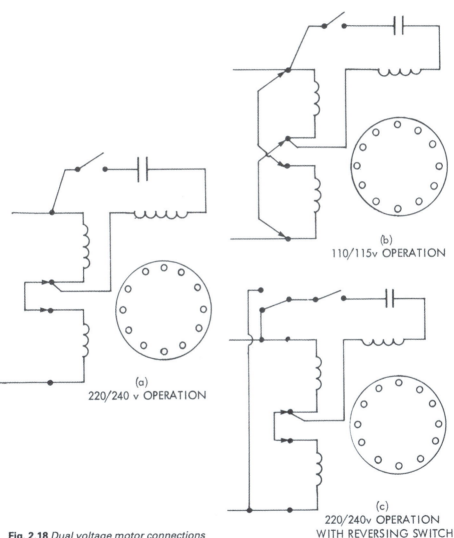

(b)
110/115v OPERATION

(a)
220/240 v OPERATION

(c)
220/240v OPERATION
WITH REVERSING SWITCH

Fig. 2.18 *Dual voltage motor connections*

usually higher. This is the starting winding. The starting winding will work correctly connected either way round – one connection will give clockwise rotation the other anticlockwise rotation. The main windings will only work correctly if the current passes through each of the windings in the same direction so that their magnetic fields add. Identification of the correct sequence by inspection is difficult so it is easier to find it by experiment. Connect the windings as in 2-18a or b and apply power for a few seconds, then reverse the connections to one of the main windings and re-apply power for a further few seconds. With the correct connection the motor will start and run normally. With one winding reversed the motor will take excessive current, will have little or no starting torque and will hum and vibrate excessively.

CHAPTER 3

Variable Frequency Drives and 3 Phase Operation

3.1 General

Motors larger than about 1/8 H.P. (100W) in industrial machinery are usually three phase machines as this avoids the complications and drawbacks of centrifugal switches and starting windings. In addition, unlike single phase induction motors, three phase machines are much easier to operate over a wide range of speeds. In most small workshops three phase supplies are not available and it is necessary to find some method of operation from single phase supplies.

The main possibilities are:-
Replacement with single phase motor
Supply from single to three phase con
verter
Individual motor conversion
Supply from Variable Frequency Drive
unit

3.2 Replacement

This involves replacing individual motors with their single phase equivalents and is probably the least attractive option. The main problem is that three phase motors are often smaller than their single phase equivalents and it may be necessary

to fit motors of lower rating and possibly with different shaft sizes and fixing centres. The one saving grace is that the majority of industrial machinery is rated for continuous full load operation, day in day out, at ambient temperatures of up to at least 40°C/105°F. Conditions in the average small workshop are nothing like as severe as this and much machinery will survive very well with not much more than half the normal horsepower installed.

Air compressors are one major exception to this comment. The motors are often used to the limit of their capacity and, unless an automatic pressure release system is fitted, place an exceptionally heavy and frequent starting load on the motor. Capacitor start single phase motors are the only type that can safely be used in this duty and should preferably be 50% larger in H.P. rating than the original three phase motor

A few special purpose installations may be upset by the larger twice supply frequency mechanical vibration which is inherent with single phase motors even when resiliently mounted. Precision surface grinders are in this category.

3.3 Supply from Single to Three Phase Converter

3.3.1 Static Converters
Commercial static single to three phase converters are readily available with maximum power ratings ranging from 2 to 20 H.P. (1.5 to 15kW)

European 50Hz models basically consist of an auto-transformer to raise the 240V input to the 415V line voltage which is used directly for two of the three line outputs. The third line output, the "phantom phase" is then fed via capacitors from one of the 415Vlines - a high value for starting with, a lower value for running. This results in an open Vee three phase system (see 1.15)

North American models usually rely on the use of 230V connected three phase motors. These do not need an auto-transformer as this voltage can be obtained directly from the line to line 230V available from cooker/power circuits.

The series capacitors feeding the phantom phase, in conjunction with the inductive reactance of the motor load, provide the phase shift needed to give a reasonable approximation to a balanced three phase output.

The term "reasonable approximation" is used advisedly as the optimum value of the phase shifting capacitor changes with the load placed on the motor, the size of the motor and whether the motor is starting or up to running speed.

The change with motor load is not too much of a problem. It is usual to choose a capacitance that is right for full load operation. At part load, because the motor impedance rises, there will be some over voltage on the capacitor phase causing increased iron and copper losses. However, at part load, the overall losses are low anyway so this extra loss is unimportant.

The change with motor rated horsepower is more serious and most converters are fitted with a switch to select the start and run capacitor size appropriate to the motor rating.

The starting capacitor is the essential component as, without this, the motor cannot start and run up to operating speed. It must not be left in circuit once the motor is up to speed or it can cause in severe overvoltage on the phantom phase. It can however be simply disconnected and not replaced by a "run " capacitor. Some converters use this low cost option. The penalty is extreme unbalance in the currents in the motor windings causing additional heating in the single fully utilised winding. The motor can still produce its full rated output for short periods but, for long term use, overheating limits the permissible output to about two thirds rated HP.

3.3.2 Starting Torque
During starting, because of the heavy currents taken by the motor and the low impedance presented to the capacitor phase, the converter three phase output is both unbalanced and below rated line voltage. Because of this the motor starting torque is reduced and may be as low as half the normal value. This is sufficient for most equipment but may well be troublesome on items such as air compressors and hoists which have a high initial starting torque requirement. A converter of the next larger power rating will reduce the voltage drop during starting

but not improve the unbalance.

If good starting torque is essential it is better to change to the rotary type of converter. This both improves starting torque and is more tolerant of multimotor loads.

3. 3.3 Rotary Converter

The static system is fine when only driving a single motor but has problems when driving something like a milling machine that may have a 3 H.P. spindle drive motor with ¼ H.P. table traverse motors.

With more than 10:1 difference in motor rating no single value of capacitance is satisfactory. The simplest way of dealing with this is to alter the machine wiring so that the traverse and coolant motors can only be operated if the main motor is running. This solves the electrical problem - the power ratio is now less than 1.1:1, but it is very undesirable from the safety point of view to have the main spindle motor running when it is not really needed.

A much better solution is to change to a rotary converter system and this method is often used for home built converter systems. This is basically a static converter with an unloaded "idler" motor connected directly to the converter output. Because this motor is running light it consumes very little power. The power rating of this motor is not critical - within reason the larger the better. A good starting point is about the same HP as the largest load motor. The idler motor gives two important benefits. Firstly it reduces the ratio between minimum and maximum total rated horsepower but it also makes it possible to maintain almost balanced three phase line voltages while other motors are starting and

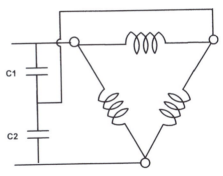

FIGURE 3.1 ROTARY CONVERTOR CAPACITOR ARRANGEMENT

running up to speed. This results in a major improvement in starting torque.

This latter behaviour is because the back EMF produced by the rotor of a three phase motor fed and running on single phase input also generates correctly phased output voltage on its unused third connection. If a large idler is used, sufficient power is available from this output to drive smaller motors at almost full continuous rating without requiring additional "run" capacitors.

Run capacitors can still improve the balance of the three phase lines but the choice of values is far less critical. As far as phase shift is concerned the load inductance is simply the inductance of the two (or more) parallel connected motors. Since the large idler has no mechanical load the system is always working at a small a fraction of its possible maximum load. Because of this part load configuration a single capacitor connected live to phantom is no longer optimum because, if it is chosen for correct phase angle, the phantom phase voltage will be excessive.

This difficulty is avoided by using two

47

capacitors arranged as a voltage divider - see figure 3.1. The effective capacitance of the two capacitors connected in series across the supply is the sum of the capacitances because the source impedance of the supply is zero and, as far as AC signals are concerned, this effectively parallels the two capacitors.

Because it also acts as a voltage divider, this sum capacitance C1+C2 is effectively fed from a voltage of supply voltage times C1/(C1+C2) where C1 is the top capacitor and C2 is connected phantom phase to neutral. This produces the required total capacitance value at the necessary reduced voltage.

Because it looks nicely symmetrical it's often thought that C1 and C2 should be equal and any inequality in their optimum value must result from some strange second order effect. This is NOT true. There is nothing magic about equal C1 andC2. It simply results in a capacitor of effective value C1+C2 fed from half the supply voltage. At this low effective supply voltage it is only possible to get close to balanced operation at no load or light loads which enable the rotor to operate close to synchronous speed. As the load increases, with consequent slowing of the rotor speed, the total capacitance needs to increase with both more in C1 and less in C2. By the time full system load is reached the optimum value for C2 is normally zero.

In home brew versions of this type of converter, provided the idler is first run up to speed without a load motor connected, it is sometimes possible to get the idler to start with a start capacitor not much larger in value than a normal run capacitor. This can get rid of the requirement to switch out the start capacitor but it's not always possible to find a capacitor value that is sufficiently large to give reliable starting without also generating excessive phantom phase voltage when run up to speed.

3.3.4 Relay and Contactor Operation

With either of the three phase converters types it is necessary to ensure that any relay or contactor coils in the machine circuits are connected to the two directly fed lines. They must not be connected to the phantom phase. This phase is subject to large variations in voltage during starting, particularly if no idler motor is connected, and may at times be less than half nominal.

3.3.5 Line to Neutral Loads

Some three phase equipment uses 4 wire input i.e. 3 line plus neutral. The neutral is not used for motor loads but may be used in a line to neutral connection for small single phase loads. Most converters have no neutral and cannot drive a line to neutral load. Fortunately this use of the neutral connection is fairly rare and the problem can be solved by disconnecting BOTH sides of the single phase load and taking them directly to the appropriate point in the input single phase power. On no account leave either wire connected to the converter output as this may result in large and unpredictable voltages being applied to the single phase load.

3.4 Direct Motor Conversion

Although motor replacement or the use of a single to three phase converter is the most straightforward way of operating three phase machinery from single phase supplies, there are many occasions when it is more flexible and more economic to convert individual motors with custom built start/run gear.

The techniques required are not difficult and the motor performance is in most cases as good or better than can be achieved with commercial converters.

3.4.1 Operating voltage

The first problem is to match the rated motor voltage to the supply voltage. Although European three phase motors are almost invariably connected for 380/420V line to line voltage, this is because the three windings are connected in a star configuration (see figure 3-2) so that the line voltage is shared between two windings in series. If the windings are reconnected in a delta configuration (figure 3-3) the whole line voltage now appears across each winding. This reduces the rated motor voltage by a factor of 1.73 (i.e.$\sqrt{3}$) to 220/240 V which can be supplied directly from normal domestic single phase power. In this connection the full load input current is increased by the same factor so that the power input and the rated horsepower remain unchanged.

Not all motors are suitable for reconnection - some types, particularly very old machines, have the start point buried in the windings with only the three line connections brought out to terminals. With sufficient determination it is usually possible to dig out the start point and reconnect the windings. The star point is usually the first joint made so it may be buried under the main leadout joints.

Rules for reconnection are very simple. Arrange the wires as three pairs of start and finish wires. Each finish wire should connect to the next start wire - the sequence in which the windings are

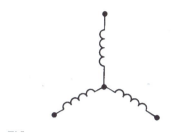

FIGURE 3.2 STAR CONFIGURATION OF MOTOR WINDINGS

connected doesn't matter so long as finish always connects to start. These joints are in the high temperature parts of the motor so they should be insulated with high temperature heat shrink sleeving or plumbers Teflon tape.

Fortunately most machines have all six ends of the windings brought out to terminals and reconnection is simply a matter of changing the links. These machines are easily recognised by their nameplates which will specify 380/420V AND 220/240V operation and also by six terminals in the connection box. Manufacturers colour code and/or allocate distinguishing letters and numbers to the six wires and terminals with what can only be described as gay abandon - perhaps I have been unlucky but I have yet to find two motor types with the same

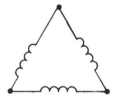

FIGURE 3.3 DELTA CONFIGURATION OF MOTOR WINDINGS

49

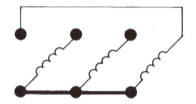

FIGURE 3.4 MOTOR TERMINALS STAR
CONNECTION

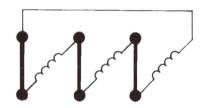

FIGURE 3.5 MOTOR TERMINALS
DELTA CONNECTION

leadout coding!

Fortunately the arrangement of the terminals and links is quite standard. Figure 3-4 shows the links in the normal 380/420 V industrial connection with the lower three terminals linked together to form the star point. Input power is normally connected to the upper three terminals. Figure 3-5 shows how to rearrange the link positions and power connections for 220/240 V

operation.

North American industrial three phase motors are commonly 230 V or 460V 60Hz line to line voltage. This 2:1 ratio is unsuitable for star/delta voltage change so dual voltage motors are provided with two sets of windings - see Figure 3.6

With Star connected motors the first set is a 230V star triplet, the second set is three individual 230V windings. These

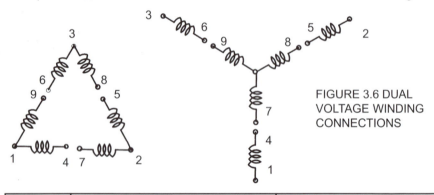

FIGURE 3.6 DUAL
VOLTAGE WINDING
CONNECTIONS

	DELTA		STAR	
HIGH VOLTAGE	LINK 4 -7, 5-8, 6-9 LINE INTO 1, 2, 3		LINK 4-7, 5-8, 6-9 LINE INTO 1, 2, 3	
LOW VOLTAGE	LINE INTO	2 - 4 - 8 LINKED 1 - 6 - 7 LINKED 3 - 5 - 9 LINKED	LINE INTO	1 –7 LINKED 2 – 8 LINKED 3 – 9 LINKED 4 - 5 - 6 LINKED

individual windings can either extend the arms of the star triplet for 460V operation or can be connected to form a second 230V star triplet paralleled to the first for 230V operation. This voltage can be obtained directly from the two live connections from domestic cooking/power circuits that are supplied with grounded centre tap 115-0-115 V 60Hz power.

Star connection is the usual arrangement for the smaller motor sizes. Large motors may be Delta connected which has the alternative winding arrangement and linkages shown in figure 3.6. This information is normally printed on or supplied with the motor. If it's not legible the types are easily distinguished by an ohmmeter check. Star winds have one triplet and three pairs of leadout wires. Delta wind results in three triplets.

With the correct single phase voltage applied to two of the three motor terminations the motor will now run as a single phase motor *provided* it is first run up to half or preferably two thirds of its rated full load speed. Because power is applied to two terminals only, two of the three windings are only partly utilised. The single fully utilised winding will reach its normal full load current when the machine is delivering about half its rated horsepower. The machine is still capable of delivering its full rated horsepower but only for short periods as this winding will then be carrying well over its rated current and will overheat.

3.4.2 Capacitor Value

Half rated power, with short bursts approaching full power, may well be enough for light duty applications. If more is needed it is necessary to add a capacitor to provide the phase shift needed to supply the phantom phase line. The optimum value of the capacitor varies with the motor design and the load that the motor is driving. It is not particularly critical - suitable values are shown in the following table:-

| RATING | | 220/240V | 220/230V |
H.P.	kW	50Hz	60Hz
0.25	0.18	10µF	8µF
0.33	0.25	13µF	11µF
0.50	0.37	20µF	16µF
0.75	0.55	30µF	24µF
1.00	0.75	40µF	32µF
1.50	1.10	60µF	48µF
2.00	1.50	80µF	64µF

If the correct value of capacitor is not available choose the next smaller value - even half the recommended value is a useful improvement. Do not exceed the values in the table - higher values will not improve full load performance and may result in excessive third phase voltage when the motor is running light.

These capacitors must be rated for 250V A.C. working or higher. Paper or polypropylene dielectric capacitors used for industrial power factor correction are very suitable - these can often be reclaimed from discarded fluorescent lamp fittings. Some capacitors will only be marked with their D.C. rating. In this case the rating should be at least 350V and preferably 500V D.C. Electrolytic capacitors are not suitable.

3.4.3 Starting Capacitors

With the "run" capacitor fitted (figure 3. 7) the motor will usually run and but the starting torque will be very low. To obtain adequate starting torque a larger capacitor must be switched in until the motor has run up to speed. Most workshop motors start

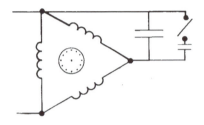

FIGURE 3.7 MOTOR WITH RUN AND
START CAPACITORS

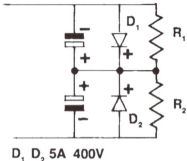

D₁ D₂ 5A 400V
R₁ R₂ 100KΩ ¹/₂ W

FIGURE 3.8 A.C. OPERATION OF
ELECTROLYTIC CAPACITORS

on light load and an additional capacitance
of twice the value shown in the above table
will be sufficient. For starting against
heavier loads, larger capacitance will be
needed - three to eight times the "run"
capacitor value is the useful range.

Because these capacitors are not in
circuit long enough for self heating to be a
problem it is possible to use electrolytic
capacitors in this position. These are very
much smaller and cheaper than "run"
capacitors of similar rating. Special
intermittent A.C. rated, motor start
capacitors are made for this duty but they
may be difficult to obtain, other than as
spares for standard motor types.

D.C. rated electrolytic capacitors are
used in enormous quantities in electronic
applications and are readily available from
electronic stockists. These cannot be used
singly because reverse voltage applied to
a D.C. rated electrolytic capacitor will
destroy it.

However, series connected in pairs
with protective diodes they make excellent
motor start capacitors.

Figure 3.8 shows the arrangement. D₁
and D₂ protect the capacitors against
reverse voltage. A₁ and A₂ discharge the
capacitors when not in use. Each capacitor

should be double the required value
because, when connected in series, the
capacitance is halved. The capacitors
should be rated for at least 350V D.C.

3.4.4 Control gear

Having sorted out the right start and run
capacitors these need to be switched in and
out at the right moment by the control gear.
While, in principle, the starting capacitor
can be switched in and out of circuit
manually it is only too easy to get it wrong
and finish up with a burnt-out motor.

Control gear for starting needs to
simultaneously apply power to the motor
and connect the starting circuit for a
sufficient time for the motor to run up to
rated speed.

It is important that power is not applied
before connection of the starting circuit - a
stalled motor typically draws five times its
rated full load current.

Since its power efficiency at stall is zero
instead of the normal 80% or so the stalled
motor is dissipating twenty five times its
rated power! Motors are designed to

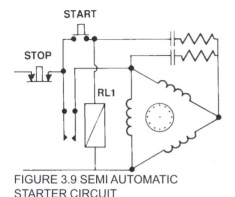

FIGURE 3.9 SEMI AUTOMATIC
STARTER CIRCUIT

withstand this sort of temporary overload but for seconds, not minutes.

It is also necessary to disconnect the starting circuits soon after the motor has reached operating speed. Most starting circuits temporarily overload the motor but since this overload is typically less than 2:1 there is much more latitude here and a delay of even half a minute is acceptable.

3.4.5 Semi-automatic Starter

A very simple semi-automatic arrangement is shown in figure 3.9. When the START button is pressed power is applied to the start capacitor and, at the same time, RL1 closes and applies power to the main motor windings. The START button should be released as soon as the motor is up to speed. This disconnects the start capacitor from the motor circuit but sufficient current now flows backwards through the capacitor to hold RL1 closed so that the motor continues to run. If the STOP button is pressed RL1 opens, disconnects power, and stays open until the next time that the start button is pressed.

A standard 220/240V A.C. relay can be used provided the contacts are heavy enough to carry the motor starting current. A push button must be used for the START switch to ensure that it is released as soon as the motor is up to speed. However, it is sometimes an advantage to replace the OFF button by an ordinary switch. This now acts as a master switch. When it is off the motor is always off and cannot be restarted even if the START button is pressed accidentally.

One difficulty that occurs with most of these "start" and "run" capacitor systems is sparking when the second capacitor is brought into circuit. If a charged capacitor is connected directly to an uncharged capacitor a very high peak current will flow until the charge equalises on both capacitors. This causes rapid wear of the relay contacts and, in extreme cases, can cause light duty contacts to weld together.

To avoid this it is advisable to connect a small resistor in series with either the "run" or the "start" capacitor. If plenty of starting torque is available connect the resistor in the start circuit - this will result in a minor decrease in starting torque but no power will be wasted in the resistor once the start button is released. If maximum starting torque is needed then connect the resistor in the run circuit - the starting torque will now be almost unaffected but there will be a small drop in motor efficiency because of the power lost in the resistor.

Use about 3 ohms for a half horsepower motor and proportionately less for higher powers. Ten watt wirewound resistors are suitable - for very low values use several resistors connected in parallel. Alternatively a foot or so of wire from an old electric fire can be pressed into service.

3.4.6 Automatic Start Systems

The above semi-automatic starting system

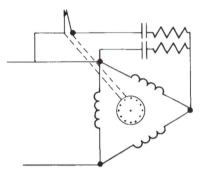

FIGURE 3.10 CENTRIFUGAL
STARTER CIRCUIT

is recommended for most workshop
applications. It is a simple system easily built
from standard components and will work with
any of the commonly available types of three
phase motors. For some items, however,
fully automatic control is essential.

One obvious system is to use an
interval timer to switch in the starting
capacitor for a fixed length of time each time
the motor is started. This is a perfectly

practical system but suffers from the
difficulty that, if an unexpected temporary
overload stalls the motor, it will stay stalled
and overheat.

To avoid this problem single phase
motors always use a speed or a current
sensor to switch in the starting windings and
these are the preferred methods for three
phase starting systems.

3.4.7. Centrifugal Starter

Speed sensing using a centrifugal starter
switch taken from a single phase motor is
a very simple arrangement - see figure
3.10.

The centrifugal element does not have
to be on the actual motor shaft, it can be
mounted separately and belt or gear driven.
It may also be more convenient to use a
microswitch instead of the original contact
points.

This is an excellent system for the
mechanical minded. Apart from choice of
start/run capacitor it is almost completely
independent of the characteristics of the
motor and the load. The fact that the
centrifugal switch is by far the most
commonly used system to start single
phase motors speaks for itself.

3.4.8 Current Relay

This system uses the high current taken
by the motor while it is running up to speed
to operate a current relay which switches
in the starting capacitor - see figure 3.11.
When the motor reaches operating speed
the current drops and the relay opens,
disconnecting the starting capacitor.

The current relay needs to be a special
design with only a small difference between
closing and opening currents (most relays
have a large differential - sometimes more

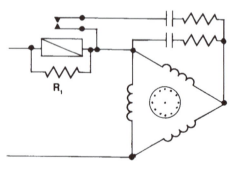

FIGURE 3.11 CURRENT RELAY
STARTER CIRCUIT

Fig 3.12 *Typical domestic refrigerator starting relays*

than 2:1). A suitable relay is the starting relay from a domestic freezer or refrigerator (figure 3-12). This relay is usually mounted on or very close to the electrical terminations of the sealed compressor unit. Sometimes a bi-metal thermal overload sensor is part of the assembly.

The contacts on the relay will normally be large enough to switch capacitors for motors up to about ¾ H.P. For larger motors or for frequent stop/ start operation the current relay should control a suitably rated second relay which does the actual switching of the capacitor.

Because refrigerator compressor motors are quite low power devices the current relay will normally be too sensitive. The sensitivity has to be decreased until the closing and drop out currents are a little greater than the full load motor current. This can be done either by rewinding the coil with fewer turns of thicker wire or by diverting the additional current through a shunt resistor (R1 in figure 3-11).

Rewinding the coil is usually fairly straightforward but not a job for a novice. The shunt resistor method is more forgiving but commercial resistors of the right value and rating may not be available. Fortunately suitable items can be readily made from length of electric fire element wire.

This method is specific to a particular motor. Some experiment will be needed to arrive at a sensitivity which will always hold the relay closed up to at least half full load speed but allow the relay to open when full speed is reached with the motor driving its maximum load.

3.4.9 Modified Time Delay Starter
A simple time delay starter system that can be built with standard components is shown in figure 3.13. This uses a double pole changeover relay to change over between separate start and run capacitors when the motor has run up to speed. This avoids the high peak currents that occurs when two charged capacitors are parallel connected

55

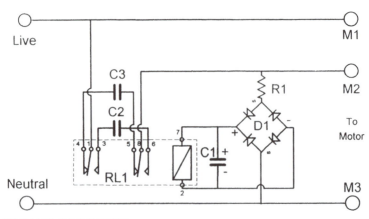

FIG 3.13 TIME DELAY STARTER

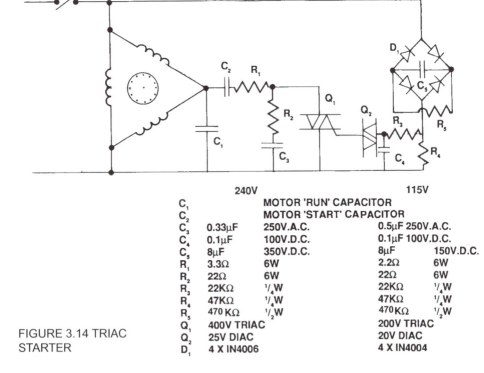

	240V		115V	
C_1		MOTOR 'RUN' CAPACITOR		
C_2		MOTOR 'START' CAPACITOR		
C_3	0.33µF	250V.A.C.	0.5µF	250V.A.C.
C_4	0.1µF	100V.D.C.	0.1µF	100V.D.C.
C_5	8µF	350V.D.C.	8µF	150V.D.C.
R_1	3.3Ω	6W	2.2Ω	6W
R_2	22Ω	6W	22Ω	6W
R_3	22KΩ	$^1/_4$W	22KΩ	$^1/_4$W
R_4	47KΩ	$^1/_4$W	47KΩ	$^1/_4$W
R_5	470 KΩ	$^1/_2$W	470KΩ	$^1/_2$W
Q_1	400V TRIAC		200V TRIAC	
Q_2	25V DIAC		20V DIAC	
D_1	4 X IN4006		4 X IN4004	

FIGURE 3.14 TRIAC STARTER

so the current limiting resistors of figure 3.9 are unnecessary. A standard double pole double throw 240V AC relay is used but, to provide the time delay, it is shunted by a capacitor and fed from series resistor and bridge rectifier for DC operation. Although the relay is rated for 240V AC operation it closes at a much lower voltage when fed with DC as in this circuit.

R1 is connected to the motor phantom phase instead of supply live because this automatically increases the switching time delay if the phantom phase voltage is low because the motor is taking longer to run up to speed

C1 and R1 are typically 10 μF 250V and 10Kohms 3W for 240V operation. About twice as much capacitance and half the resistance for 115V components.

C2 and C3 are respectively the appropriate run and start capacitors

3.4.10 Triac Starter

This one is for the electronic buffs - its main advantage is that the components are readily available at most electronic stockists.

The circuit arrangement is shown in figure 3.14. When power is first applied, current flows through A_4 charging up C_5. The voltage drop across A_4 switches on Triac Q1 via the Diac Q_2. This brings into circuit the "start" capacitor C_2. After about one second C_5 is so nearly fully charged that the reduced charging current is insufficient to trigger Q_1 and Q_2 so that Q_1 reverts to its "off" state disconnecting the "start" capacitor.

When the motor is switched off C_5 discharges through A_5 leaving it ready to time the next start cycle.

3.4.11 System protection

Three phase motors should never be overloaded to the point at which they stall or they will rapidly overheat. When operating from a true three phase source the motor will safely and immediately restart if a short term mechanical overload ceases. This is not true of many of the static phase converter systems and the motor may remain stalled and overheat. Fuses alone are not adequate protection because a fuse that will accept the starting current peak will not necessarily blow with a stalled motor. Protection devices that deal with this problem are covered in section 8.5

3.5 Supply from Variable Frequency Drive unit (VFD)

For home workshop use, particularly when variable speed is an attractive attribute, VFDs are generally the most flexible and useful method of single to three phase conversion.

They are sometimes called Inverters. While this correctly describes the device the name is less specific. Inverters may be fixed frequency devices without provision for variable frequency operation.

A typical VFD is shown in figure 3.15. This is a small 0.4Kw VFD suitable for driving a small lathe or drill press. VFDs are rated by the mechanical power output of the largest motor they are capable of driving. The powers mostly line up pretty well with the common motor HP rating. 1 Kw mechanical power corresponds to 1.34 HP so that this 0.4 KW VFD is capable of driving a ½ HP motor.

3.5.1 Basic operation

The basic essentials of a VFD are shown in figure 3.16 The power input, dependent on VFD type, may be single phase or three phase. It is first rectified and then smoothed by a large capacitor to provide a high voltage DC link. Three sets of semiconductors then rapidly alternate the motor connections between the + and - ends of the DC link in a sequence controlled by a central microprocessor. This switching frequency (usually called the CARRIER or PULSE FREQUENCY) is typically in the range 2 to 15KHz so that many switched pulses make up a single cycle of the VFD output to the motor.

The semiconductors act as true switches being either fully " ON" or fully "OFF". This means that the DC link switched output can only be a train of rectangular pulses. This is converted into the three phase set of sinusoidal currents required by the motor by control of the "ON" to "OFF" ratio of the appropriate parts of the rectangular pulse train. During each "ON" period the current starts to build up in the inductance of the motor winding but before it can reach its maximum value it is switched off by the ending of the drive

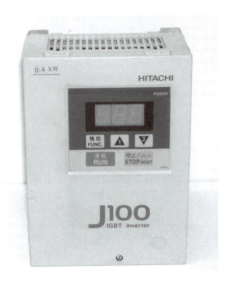

Fig 3.15 *A typical VFD*

pulse. The motor responds to the average value of the series of current pulses built up during each cycle of the VFD output

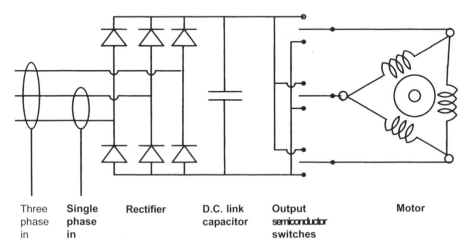

| Three phase in | Single phase in | Rectifier | D.C. link capacitor | Output semiconductor switches | Motor |

FIGURE 3.16 BASIC VFD SCHEMATIC

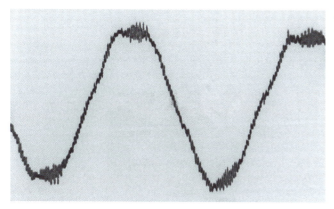

Fig 3.17 *VFD output current waveform*

frequency and this can be seen in the current waveform shown in figure 3.17 .

This is a single shot picture showing the waveform at one instant in time. It shows the current wiggles resulting from the pulsed input and the distributed inductance and capacitance of the motor windings.. Because there is usually no simple synchronous relation between the carrier frequency and the three phase output frequency, the different phasing results in slightly different wiggles in each successive cycle of the three phase output frequency. Over many cycles this waveform shape averages out to an acceptable approximation of the desired sinusoidal current required by the motor.

Although the voltage waveform reaching the motor shown in figure 3.18 (also a single shot photo) is basically a rectangular wavetrain it is strongly modified by high frequency components arising from the inductance and capacitance of the motor windings. These high frequency voltage variations have little effect on motor torque so the waveform is acceptable. The transient voltage overshoots, however increase the voltage stress on the motor

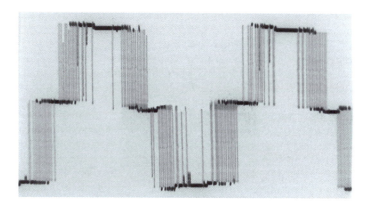

Fig 3.18 *VFD output voltage waveform*

59

insulation and slightly increase the motor power loss. This needs to be taken into account in the choice of the motor.

It's perhaps worth emphasizing that these slightly surprising waveshapes are not oddities that may occur on a particular VFD but are typical of normal VFD output waveforms.

3.5.2 Voltage /frequency relationship

As explained in section 2.4 standard induction motors operate at full load speeds of typically 50 to 100 rpm less than synchronous speed. In the fixed supply frequency case they operate at nameplate supply frequency and voltage. In the VFD case, when the internal oscillator sets a new synchronous frequency, the nameplate supply voltage is no longer correct and the VFD has to supply a new voltage appropriate to that speed. What is needed is to always operate the motor at its design flux density. With a perfect loss free motor this results in a linear relationship between supply frequency and the required supply voltage i.e. half speed needs half nameplate frequency and half nameplate supply voltage.

VFDs are organised to do this, and this mode of operation is usually called the V/F or V/Hz mode. Real motors are not perfect and, while the imperfections are of minor importance at rated supply voltage and frequency, they become more serious when operated far from the original design values.

The first problem is winding resistance - only a few % of the supply voltage is lost in this resistance in normal operation but, at perhaps 10% of rated speed, the voltage drop becomes ten times more important and requires a corrected voltage profile.

The same applies to the speed regulation. A 75 RPM drop from no load to full load is pretty acceptable at, for example, rated 1425RPM speed but is excessive at low demanded speeds.

Most VFDs include a special program that monitors the resistive and reactive component of the motor currents and derives from this a corrected V/F relation based on the average winding resistance and speed regulation values of a typical motor. Names vary a bit but something like Sensorless or Flux Vector Control is a recognisable alternative to V/F control. Compensation is more than good enough for most applications but, because Flux Vector control is based on average motor parameters, it may need further adjustment to optimise it for a particular motor. Some VFDs have a special installation routine that measures the actual parameters of the installed motor. This input is then used to correct the V/F relationship.

The factory default setting will normally be for a VFD driving its largest permissible motor power. While it's OK to drive any smaller size motor, smaller motors will be under corrected with poorer speed regulation and some degradation of low speed torque.

V/F and Flux Vector control programs are "open loop" programs i.e. they provide the input conditions that should achieve the desired speed under the average range of load conditions but they do not ensure that the achieved speed is precisely equal to the programmed speed. For applications where precise speed control is essential some VFDs also include facilities for "closed loop" control. For this mode a tacho generator is mechanically coupled to the load motor. The tacho generator provides a signal directly proportional to motor speed and the control system adjusts the motor drive until this tacho speed signal equals the programmed speed .

Although this mode can achieve precise control of speed it needs careful setting up for proper operation. The adjustments provided to achieve this are usually arranged as PID control signals – the Proportional, Integral and Derivative components of the error signal.

3.5.3 Data entry

VFDs are provided with a wide range of user selectable control functions and these can be used to trim the performance with a particular motor. Some VFDs provide a selection of switches and potentiometers to control the functions but in most of the larger units data entry is by a keyboard controlling a cascading menu system. Each manufacturer uses his own particular arrangement of menus so it is essential to have access to the instruction manual. This is no problem with new equipment but may be a major stumbling block with second hand items. Fortunately major manufactures operate their own Internet Website and most of them allow you to download the full text of the instruction manual.

Most switch and keyboard entry systems change the inputs to the microprocessor rather than controlling the function directly. The microprocessor only reads the revised inputs during power up so that it is only after a subsequent power up or by operation of a separate data entry key that the changes become effective.

Many VFDs offer a bewilderingly large number of user definable options – as many as 100 is common. Don't be overawed by this plethora of possibilities. For straightforward small workshop use practically all the options should be left unchanged at their factory set default values. Perhaps the most useful command in the instruction set is the one that returns all settings to factory default values!

The essential information that has to be entered by the user is the motor nameplate information that is used by the V/F and Flux vector control programs for their input data. Most key entry VFDs have their own keyboard and display to enter this information but a few industrial types intended for permanent installations have no built in keyboard and this has to be purchased separately.

3.5.4 Speed considerations

With the nameplate data safely entered the VFD can speed control the motor from zero up to its nameplate full load speed. The motor can only deliver its full rated HP when operating at its normal nameplate speed. At any lower speed it can only deliver the same torque which means that the available HP drops in the same ratio as the speed. This is suitable for fans and similar devices because their HP requirements drop as the speed drops. Not so good for many machine tool types which need full or even increased HP at lower speeds. When choosing a VFD and motor type it is essential to consider the low speed behaviour and size the motor accordingly. To get the full benefit of replacing a single speed motor setup with a VFD system it may be necessary to fit a considerably higher power motor.

This is fine when full power low speed use is occasionally needed and for minutes rather than hours but, for continuous full power low speed use, it is necessary to fit an external fan to cool the motor because the internal fan cannot move sufficient air at these low speeds. This is an oft quoted limitation of low speed VFD use and manufacturers quite rightly emphasise it in their literature. However in non - professional installations, continuous operation at maximum ambient temperature, full load and low speed is pretty rare event so external fans are commonly omitted. If tempted to touch the motor casing to check the temperature rise be careful - modern motors while running well within their ratings can still be hot enough to raise a nasty burn on an incautious finger!

Different conditions apply when the VFD is set to call for speeds higher than the motor nameplate speed. The squirrel cage rotors of induction motors are pretty rugged devices and normally capable of running at as much as twice normal speed with little penalty apart from slightly reduced bearing life and higher noise level. The difficulty is the V/F law and the fact that motor losses increase at higher supply frequencies.

Operation at twice the speed implies supplying twice the supply frequency at twice the supply voltage. However standard VFDs cannot deliver an output voltage that is higher than their input voltage so, for speeds above motor nameplate speed, the VFD can only output the normal supply voltage which means that the motor can no longer deliver its rated full load torque at these higher speeds. With a perfect motor the available torque would fall linearly with demanded speed so that HP would remain constant. With real motors the losses increase with supply frequency and the torque falls off more rapidly.

The practicable maximum is dependent on both the motor design and the load that you put on it. At least one and a half times nameplate speed at half full load torque is a reasonable expectation. With an efficient motor and reduced torque more than double speed may be possible. Don't forget that, at these increased speeds, you may be exceeding the design speed of both the motor and any machinery that it is driving. Be very sure that any failure is not a safety hazard!

In normal VFD use, rated full load torque is only available up to the motor nameplate speed – higher speeds are possible but only at reduced torque. A way round this is to use a high voltage VFD to drive a lower voltage motor. For example, if a 240V 50Hz delta connected motor is driven from a 415Vin/415V out inverter, the built in V/F law allows full torque to be maintained well beyond the usual 240V limit all the way up to 415V .

If the VFD is programmed to drive a standard 240V 50Hz motor it will provide the motor with its normal correct inputs all the way up to nameplate speed but it will now also be capable of delivering full rated torque up to about 1.7x speed. There will be little change in motor current but, because this is now at 1.7x normal voltage, it is consuming nearly double its normal power and the VFD must at least be rated to drive a correspondingly higher power motor.

The motor is now delivering considerably more than its nameplate HP but it is still operating correctly because the VFD is automatically maintaining the right relation between voltage and speed. For this fairly modest speed increase there is little change in motor efficiency but the increased power input means that it will overheat if operated for long periods with simultaneous high torque load and high speed.

This trick can also be used if a lightly loaded very high speed motor is needed.

The extra voltage makes it possible to reach speeds unobtainable with 240V in 240V out VFDs. Be VERY careful if tempted to experiment in this direction – there is a LOT of energy stored in a heavy rotor spinning at 10,000 RPM and a mechanical failure can be catastrophic. Be also aware that the mechanical balance of a rotor that is designed to run at 1500 RPM may be quite inadequate at these speeds – 8:1 in speed is 64:1 in unbalance force! Windage and fan power get out of hand. The power absorbed by a fan rises as the cube of the speed so 8:1 in speed increases the power absorbed by the fan by over 500 times!!

The VFD will normally arrive with the manufacturers default settings correct for your local supply frequency – usually 50 or 60Hz. The electronics of the VFD decide the appropriate output voltage based on the ratio between your demanded frequency and this frequency - normally the nameplate frequency of your motor (in VFD terms the this frequency is referred to as the MAXIMUM VOLTAGE FREQUENCY or BASE FREQUENCY).

Although 50Hz (Europe), 60Hz (North America) are the common standards most VFDs offer a free choice of base frequency which can be set in the range from a few Hz up to about 400Hz. This can be of use in setting up the VFD for non standard motor applications.

3.5.5 Speed setting

A multiplicity of methods is available for external control of the speed setting with also provision for preset minimum and maximum speeds. The 0-100% speed can usually be set by a voltage, 0-10V, by a current, 0-20mA, 4-20mA or by a potentiometer. The external potentiometer is the most generally convenient method, the value is not critical 3 to 10Kohm is typical. Unless otherwise specified, Linear Law pots (constant resistance change per unit angle change) are often supplied. These are OK but the Linear Law results in much of the useful low speed range being squashed into a narrow angular range close to zero. A Log Law pot overcomes this by opening up the scaling at the lower end of the speed coverage.

For some applications a small set of fixed speeds is more convenient. Most VFDs offer the facility of cycling between a number of preset fixed speeds. This is fine for repetitive production operations but for general purpose use a switch selected resistor chain is more convenient. Table 3.19 shows suitable resistor chains. The first column is a six step chain 50% increase per step. The second column is a 12 step chain 20% increase per step. All resistors are standard values and readily obtainable in 1/10w or 1/4w sizes.

3.19 EQUAL INCREMENT RESISTOR CHAIN

330Ω	100Ω
470Ω	120Ω
680Ω	150Ω
100Ω	180Ω
1500Ω	220Ω
2200Ω	270Ω
	330Ω
	390Ω
	470Ω
	560Ω
	680Ω
	820Ω

3.5.6 Carrier Frequency

A number of selectable frequencies in the range 2 to 20 KHz are usually provided.

The manufacturers chosen default frequency for standard operation is typically in the range 5 to 10KHz and this should not normally need to be altered.

Frequencies at the high end of this range decrease the audible noise from the motor but slightly increase the power dissipation in both the VFD and the motor. Because of this, some VFDs specify some power derating if carrier frequencies higher than 10KHz are used.

It may be necessary to reduce the carrier frequency if there are very long leads (> 30m) between the VFD and its motor – this will usually be specified by the manufacturer.

The highest carrier frequencies are useful if the VFD is used for three phase outputs well above the normal 50 to 80 Hz range. This is to ensure that there are a sufficient number of carrier pulses in each cycle of the output waveform to permit good synthesis of the required sinusoidal current waveform.

3.5.7 Load Limitations

It's perhaps natural to think of a VFD as a sort of portable local three phase supply to which motors can be connected as required. Unfortunately this may not be true. The electronics and protection systems of the VFD are based on the currents and voltages encountered in the normal start and run sequence of the motor load. However at the actual instant connection or disconnection of a motor load the current /voltage relationships are abnormal and transient voltages and currents occur which can overstress the VFD. Because of this, most manufacturers specify that the motor be permanently connected to the VFD output terminals and that motor on/off

control is always by the VFD which has built in soft start and soft off functions. A breaker is permitted between the VFD and the input power but it is emphasised that this is for emergency use only. It would normally only be used to isolate the whole system after the VFD has powered down.

In the motor "off" state the power consumption of the VFD is extremely small and it can be left connected to the main power source indefinitely.

The possible overstress occurs when power is suddenly removed or if a second motor is switched in or out of circuit on an already running VFD. In industrial use this limitation is rarely a problem but for a small workshop perhaps powering a small mill which may be fitted with separate drive, traverse and coolant three phase motors it's a real nuisance. Multiple separate VFDs overcomes the difficulty but is expensive. The usual solution is to reserve the VFD for the main drive motor and use the methods of 3.4 or 3.5 to drive the others.

Some leeway is possible for the adventurously minded user. Study the manufacturers literature carefully. Permanent connection of a single motor is the recommendation. Multiple motors and/ or direct switching, if forbidden, are not indicated as instant failure hazards but "may increase the failure rate" or "reduce the operating life". These are entirely reasonable reservations on the part of the manufacturer because they have no knowledge of the exact type of motor and load switching to be encountered. The fact that instant failure is not predicted is a pointer that the design safety margins are sufficient to cover normal switching events but may be stretched by unusual motor loads switched at the most unfortunate phase angle and load/speed condition. One manufacturer makes the comment "If it is unavoidable to turn the main power supply on or off to start or stop the inverter, it must not exceed once per hour"

It is clearly bad practice but, if there is no reasonable alternative, direct switching can be tried and in many cases the VFD will suffer no more than a small increase in the failure rate (VFD failure rates are normally measured in thousands of hours). This is NOT a recommendation to try this - you are disregarding manufacturers instructions and if something goes wrong you will receive no sympathy from anyone - least of all the manufacturer! The type of failure to be expected is damage to some or all of the output semiconductors - a repairable but expensive fault.

Some decrease in the severity of the switching transients can be obtained by permanently connecting a resistive load in delta across the output of the VFD. This provides some damping of the voltage transients and also ensures that the VFD never sees an open circuit load. Three resistors consuming 5 to 10% rated load current is a suitable arrangement. The load switched must be well within the VFD power rating – if anything like full rated load is switched the transient peak current may activate the VFD overcurrent protection circuits which will power down the VFD..

3.5.8 Acceleration and Braking

VFDs provide a variety of selectable acceleration and regenerative braking profiles based on time, and or, torque patterns. The acceleration pattern can be freely chosen but more care is needed with the braking pattern. The trouble is that regenerative braking requires dissipation of the kinetic energy stored the inertia of the motor and load. The VFD transforms this into electrical energy and via its switched output semiconductors feeds it back into the DC link. This raises the voltage of the DC link and, if this voltage increase is excessive, safety circuits terminate the braking signal and the load coasts to a standstill in its own time.

Most VFDs can exert a braking force of 25% of full load torque without triggering the safety termination. This can be increased to about 125% of full load by adding an externally fitted high wattage braking resistor to limit the voltage rise. These are listed as optional accessories. These can make a big difference to the spindown time at the upper end of the speed range and are a very useful fit for frequent start/stop operation.

Any suitable high power resistor can be used provided it is not lower in ohmic value than the manufacturers recommended item – higher values are OK but provide less braking. The resistor has to withstand the full supply voltage for a second or so as the motor coasts to a standstill. Electric fire elements are OK but lamps are NOT suitable because their cold resistance is only about 1/6 th of the hot value. The manufacturers chosen minimum value does not vary directly with motor power because it is controlled by the rating they choose for the separate semiconductor that switches it into circuit. With 200v class VFDs, 50 ohms is typical for up to 1 HP dropping to 20 ohms at 10HP. Double these values for 400V class.

In addition to this regenerative braking, some manufactures also provide DC injection braking. This switches a unidirectional current through one winding during the braking phase and the induced currents in the rotor generate additional braking force.

3.5.9 Motor Selection

Although in principle a VFD can drive any of the standard three phase motor types (2 pole to 8 pole) it places more stress on them than in standard fixed frequency operation.

The most important difference is voltage stress. Even in the most favourable case the peak value of the VFD voltage waveform may reach peak values of nearly twice supply voltage and switching transients or long VFD to motor connecting leads can make this worse. In addition the fast edges of the voltage waveform concentrate much of the increased stress on the insulation of the first few turns of each winding so that wire to wire insulation is almost as fully stressed as wire to case insulation. Fortunately most standard industrial motors have conservatively rated insulation and stand up pretty well to this usage. Nevertheless, if you're buying a new motor, get a VFD rated motor or at least get confirmation that it's suitable for VFD use.

The peculiarly shaped drive waveform slightly increases the motor losses and motor heating but this is a fairly minor effect. As long as you're reasonably generous with motor rating it should not be a problem..

On rare occasions an existing motor installation may have power factor correction capacitors connected across the motor. These MUST be removed before connecting to a VFD.

3.5.10 Installation

The VFD manual should give guidance on wire sizes and fuse/breaker protection. This protection is rated primarily for protection of the wiring. Protection for motor overload or overcurrent is by the much more sophisticated control system within the VFD.

Control wiring and motor power wiring should be kept well apart and any long lengths of control wiring should preferably be screened or within a grounded metal enclosure to prevent motor power currents getting into the control circuits. Screening on the control wires should be connected to ground only at the VFD end of the cable run. The motor power wiring should not preferably exceed a few feet (2 or 3 metres) and, for most of its length, be as a closely run or twisted triplet of wires.

If the motor power wiring is longer than about thirty feet (10 M) a new problem arises because of the fast edges of the drive waveform. The long wires start to act as a transmission line mismatched at both ends. The reflected waves can build up excessive peak voltages - in bad cases as much as twice normal. The solution is to fit a small amount of series inductance in each lead, close to the VFD terminals. This slows the VFD waveform edges sufficiently to avoid the problem. Suitable items will be listed as accessories by the VFD manufacturer. It may also be necessary to avoid using high carrier frequencies.

VFDs can be a source of radio interference particularly in the radio Long and Medium wavebands. Manufacturers approach to this varies – some build in the necessary suppression components others supply externally mounted filter units. These filter units can be quite complex. In addition to input and output filtering they may contain circuits to protect the VFD against possible supply voltage transients and excessive peak currents.

In all cases good grounding of all metalwork is essential. Be particularly careful of any VFD external metalwork - in many cases the RFI (Radio Frequency Interference) requirements result in high capacitative currents, supply live to case. An ungrounded VFD can lead to a nasty shock.

In the case of adding a VFD to an existing milling machine or lathe installation there is a problem on what to do with the existing power switching. Although it's possible to rewire the switching and push buttons to the VFD control lines it's much safer to leave as much as possible of this undisturbed and mount the VFD Forward, Reverse, Stop and Speed controls in a separate little control box. If this is located in a hinged mount above the existing power switching, it's pretty easy to arrange that it's impossible to switch the power on or off unless the VFD is safely in it's 'motor off' state.

3.5.11 Power supply considerations

VFDs are not particularly fussy on their input supply voltage. Within reason, they will accept the local supply voltage and be capable of operating normally up to the motor speeds where the voltage required by the V/F law exceeds the supply voltage. Higher speeds are possible but, as with normal operation, the available torque will then drop with increasing speed. Because of this VFDs are often supplied as 200V class or 400V class instead of for defined particular supply voltages. The 200v class is OK for 200 to 240V nominal (+/- 10%) supplies. The 400V for 380 to 480V nominal supplies. Because they can also accept 50 or 60Hz supplies they are then usable without modification on both European and North American type power systems.

VFDs are also supplied a single phase or three phase input power versions. The output stages are identical, the difference is single phase full wave or three phase full wave rectification of the input power, and the size and rating of the DC link reservoir capacitor(s). Three phase input VFDs can still work OK, but possibly at reduced output power, if supplied with single phase input to two of its input lines.

The main three phase rectifier is normally capable of supplying the full DC link power with either three or single phase input. The difficulty is in the energy storage capacity and ripple current rating of the DC link capacitor(s). Because three phase input refreshes the charge on the DC link capacitor three times as often as single phase input, the lower single phase refresh rate increases both the ripple voltage and ripple current. Most manufacturers choose a generous value for this capacitor (the usual capacitance tolerance is -20%/+100%!) which also has to handle the carrier ripple current generated by the output stage. The increased ripple voltage is not normally a problem but, at full load, and maximum ambient temperature, the higher ripple current may exceed the capacitor ripple current rating and this will result in reduced life.

Manufacturer's approach to this varies - some permit it, some expressly forbid it although appropriate derating would seem to be a more logical stance. If you're tempted to try this, connect supply neutral and supply high to two of the three phase inputs. On older VFDs one pair of these inputs may also provide single phase power for the low voltage logic circuits. With this arrangement any pair will energise the DC link for the output stage but the VFD may

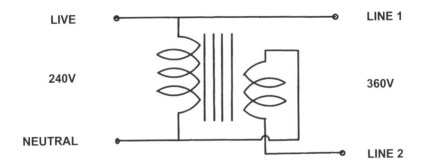

LIVE ●————————————○ LINE 1

240V

360V

NEUTRAL ●————

LINE 2

FIG 3.20 AUTOTRANSFORMER CONNECTIONS

not switch on unless you have chosen the pair that also supplies power to the logic circuits. The single phase power must of course be the same voltage as the line to line voltage of the normal three phase input.

Modern VFDs mostly use switchmode logic power supplies fed from the DC link and these are more tolerant to reduced supply voltage. Some 400V class VFDs are even capable of continuing to operate even down to a 240V in, 240V out mode.

With European 50Hz supplies it is fairly common to find machines fitted with old 415V three phase motors which cannot easily be reconnected in delta for 240V operation. 240V in, 415V out VFDs are possible but they are pretty rare and about twice the price of the standard 240V in, 240V out types.

Provided a reduction in power is acceptable, 415V motors can still be run from a 240V VFD by choice of a non standard base frequency (see3.3.2). The V/F relation shows that the 240V supply to a 415V motor needs the base frequency to be dropped in the ratio 240/415 x 50Hz = 29Hz . The VFD now thinks it's driving a 240V 29Hz motor and adjusts it's V/F law

accordingly. The motor now behaves as a 29Hz design delivering its original full load torque but at about 60% of its nameplate speed i.e. 60% of its nameplate HP.

Much of this power loss can be recovered, if you have a 400V class VFD, by driving it from an autotransformer to increase the single phase supply voltage to at or near 415V. Double wound 240V to 120V transformers for portable power tools are common items. Rearranged as an autotransformer, with the 120V secondary reconnected to add to the 240V input, 360V is available. The V/F base frequency correction is now only 360/415= 0.87 x 50Hz so that nearly 90%of the original rated speed and HP is available.

The transformer needs to be generously rated because the peaky nature of the VFD input current results in additional transformer heating. As a rule of thumb, the transformer rated secondary current should be at least 50% higher than the VFD full load input current.

Before reconnecting the secondary to add the additional voltage, be sure to check that it is truly isolated and not already connected to ground or case at any point –

some have the secondary centre tapped and the centre tap connected to ground!.

It's possible to add the secondary to the 240V live or to the 240 neutral connection to produce the 380V total. However The addition to the neutral connection as shown in figure 3.20 should be chosen as this keeps the maximum line to ground voltage at, or below, 240V.

3.5.12 Final Note

Please remember that these notes are intended to help you understand typical VFD operation and usage. VFD types can vary considerably so be sure to take account of the particular manufacturer's instructions and recommendations.

VFD drive makes it possible to operate both motors and the driven machinery at speeds well above their original design speeds.

Remember that this may invalidate any guarantee and can also be a safety hazard.

Don't forget that the DC link capacitor retains a dangerous charge for several minutes after all power has been removed. If you need to make internal adjustments, allow a delay of at least five minutes for this charge to decay to a safe level.

CHAPTER 4

Commutator Motors

4.1 Commutator Operation

The induction motors discussed in the previous chapters were able to use an alternating current supply to generate the necessary relative motion of a magnetic field to provide output torque. However, if the supply is direct current it is necessary to introduce some form of switching in the rotor or stator circuits to provide the necessary changes in field direction.

The commutator motor overcomes this problem by feeding current to windings on the rotor via carbon or metal brushes which bear on metal segments on the rotor connected to taps on the rotor windings. The metal segment assembly is the commutator and the complete assembly comprising rotor, windings and commutator is referred to as the armature. A selection of armatures is shown in figure 4-1.

Fig. 4.1 *A variety of typical armatures*

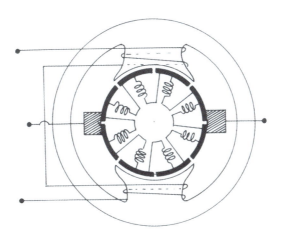

FIGURE 4.2   COMMUTATOR MOTOR

The winding arrangement is shown diagrammatically in figure 4.2. A permanent magnet, or field coil windings on an iron stator, provide a strong magnetic field at right angles to the rotational axis of the armature. Each of the armature coils is wound in a pair of diametrically opposite slots in the laminated iron rotor and the coil ends connected to adjacent commutator segments. With this connection the current flow through the brushes divides into two equal parallel paths through the windings. This results in the magnetic fields generated by the current flow in the conductors all adding across a single diameter at right angles to the external unidirectional field and this produces maximum torque on the rotor. As the rotor rotates the torque first starts to fall as the conductors move past the optimum position but it is restored as the next pair of commutator segments pass under the brushes and bring the next set of conductors into circuit. This results in almost constant torque which is only slightly dependent on rotor position.

4.2 Permanent Magnet and Shunt Wound Motors

In these motor types the armature rotates in the fixed (i.e. constant strength and position) transverse magnetic field provided by the stator.

Because the armature conductors are moving through a stationary magnetic field a voltage is generated in these conductors (the "back E.M.F.") which opposes the voltage applied to the brushes. In a perfect lossless motor running with no shaft load the speed would rise until the back E.M.F. equalled the applied voltage and the motor would only draw current from the supply while it was accelerating the inertia of the armature up to the no-load speed.

In a real motor, sufficient armature current will need to flow to overcome mechanical windage and friction losses and this current flowing through the resistance of the brushes and armature

71

windings causes a voltage drop which subtracts from the applied voltage and results in a correspondingly slightly lower no-load speed. As a mechanical load is applied to the motor the armature current will rise further to produce the necessary output torque with a corresponding further drop in speed.

It is worth emphasising that these speed changes only result from imperfections in the motor. A motor of this type with constant fixed field (e.g. permanent magnet or an electromagnet with the windings connected directly to a fixed voltage) is essentially a constant speed machine with the output speed directly proportional to supply voltage and the armature current directly proportional to load torque.

4.3 Series Wound Motors

An alternative arrangement is the series wound motor (figure 4- 3). Here the field coils are connected in series with the armature so that the strength of the field they generate in the stator is directly proportional to armature current. This

FIGURE 4.3   SERIES
WOUND MOTOR

type of motor is very useful when a large starting torque is needed. At start up the whole of the large armature current flows through the field coils giving simultaneous maximum field strength and maximum armature current. As the motor speed increases armature current drops and with it the strength of the magnetic field generated by the series field coils. This drop in field strength results in a further increase in speed. In theory, if no other load is placed on the motor, this could result in a runaway condition with the speed of the motor increasing without limit. In practice, because each increase in speed reduces the available torque, the speed rises until limited by the motor mechanical and electrical losses.

The losses in most small series wound motors will limit the speed to a safe value even if operated on no load. However, some motors are capable of no load speeds high enough for centrifugal forces to cause catastrophic failure. Unfortunately there is no easy way to discover this in advance and the only safe course of action is to ensure that there is always sufficient shaft load on a series wound motor to limit the speed to a safe value.

The series wound motor does not have the constant speed characteristic of a shunt wound machine. If a load is placed on the shaft of a series wound motor, the current which flows through the armature and field to generate the necessary torque increases the field strength. This means that if the armature is to generate the same back E.M.F. it must rotate at a lower speed. In an ideal series wound motor the shaft speed is directly proportional to supply voltage and inversely proportional to current. Because torque is proportional to armature current x field current a 4:1 increase in torque load will be needed to double the current drawn from the supply and this will result in a 2:1 drop in speed.

In real motors, mechanical and electrical losses reduce the speed increase at light load but result in even larger speed reductions when heavy loads are applied.

4.4 Universal Motors

The series wound commutator motor is not limited to D.C. operation but is also capable of being operated from A.C. supplies. Motors of this type suitable for operation from both A.C. and D.C. supplies are known as universal motors.

Although on A.C. the direction of the motor current is changing at the supply frequency, the direction of the flow of current in both the armature conductors and the field conductors reverses at the same time. Because of this the direction of the torque exerted on the armature is unchanged. The current flow varies between zero and a maximum value (positive or negative) twice per cycle of the supply frequency and this means that the torque exerted on the armature is also pulsating at twice the supply frequency. This torque variation is smoothed out by the mechanical inertia of the rotating armature and is not normally noticeable. However, as with induction motor stators (section 2.3), it is necessary to construct the stator from iron laminations to minimise eddy current losses arising from the alternating current flowing through the field windings.

Shunt wound motors cannot operate in this way because, although the voltage on the armature and field windings reverses at the same time, the high inductance of the multi-turn field winding causes the field current to lag behind the armature current so that the current reversals no longer occur at the same time. The high inductance of the field winding also drastically reduces the field current and the combined result is that shunt wound motors perform extremely poorly or not at all on A.C. supplies.

In the series wound motor the series connection ensures that the armature and field currents are in phase (it is the current, not the voltage, that controls the strength and position of the magnetic field) and this is not affected by the inductance of the field windings. However, the inductance of the field windings is an additional impedance in series with the supply which reduces the effective supply voltage when the motor is taking a large current. This results in a small reduction of the maximum torque that the motor can produce when operated from an A.C. supply. Most applications can tolerate this slight loss of performance but occasionally, when the last ounce of performance is needed, the A.C. supply is first rectified with a full wave bridge rectifier so that the motor is in fact operated from a D.C. supply. Permanent magnet and shunt wound motors can also be operated in this way and it is a common method of operating low voltage commutator motors when the constant speed characteristic of a shunt wound motor is needed.

4.5 Compound Wound Motors

When the supply voltage is first applied to a shunt wound motor the inductance of the field coils delays the build-up of current and in the case of a large motor it may be as long as a second before the field current reaches its normal value.

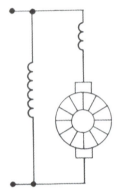

FIGURE 4.4 COMPOUND WOUND MOTOR

At the instant of switching the armature is stationary and cannot develop back E.M.F. so it draws a heavy current from the supply which is limited only by the resistance of the armature conductors. A large efficient motor may try to draw twenty times its normal full load current and the current will remain excessive until the field current builds up towards its normal value and allows the motor to accelerate to its rated speed.

This large current surge causes difficulties in the motor switching and protection circuits. In extreme cases it can result in such a large voltage drop in the supply leads that the voltage actually reaching the field terminals is not high enough to allow the motor to generate enough starting torque and the motor will fail to start.

The series wound motor does not suffer from this problem as the initial current surge passes *more* current through the field winding and increases the available starting torque.

The compound wound motor is basically a shunt wound motor with a small additional series field winding wound in with the shunt field to improve the starting characteristics (figure 4-4). In a typical compound wound motor, at full rated speed and load, 90% of the field ampere turns are provided by the shunt winding and the remaining 10% by the series windings. In the critical initial start up period the series windings generate almost all the field ampere turns providing a total field roughly equal to the normal shunt field.

The series winding is connected so that the current through it aids the shunt field. This means when a load is placed on the motor the increased armature current also increases the strength of the field with a consequent drop in shaft speed. In the example quoted above the additional drop in speed, no load to full load, due to the series compounding windings will be rather less than 10% so that the constant speed characteristic is not much worse than that of a pure shunt wound machine.

4.6 Motor Reversing

The shaft rotation of series or shunt wound motors is independent of supply polarity but it can be reversed without difficulty by reversing the connections to either the brushes or to the field coils. Some high speed motors (e.g. vacuum cleaner motors) have the position of the brushes slightly skewed in the direction of rotation to improve the commutation and will be slightly less efficient when running in the reverse direction. However, this is unusual, and most motors have their brushes in the central neutral position and will run equally well in either direction.

Compound wound motors can be reversed by changing over the brush connections. The alternative method of reversal by changing the field connections is more difficult. The series field must always aid the shunt field so if this method is used *both* the series field and the shunt field connections must be reversed.

Permanent magnet motors can be reversed by interchanging the brush connections and this is of course exactly equivalent to reversing the supply polarity.

It is often necessary to be able to reverse a motor with the minimum of external wiring and switching - remote control actuators are typical of this type of requirement. The most popular arrangement is the "split series motor". This is a series wound motor fitted with two series field windings – one connected for clockwise rotation and the other for anti-clockwise (figure 4-5). Only three wires are needed to the motor and reversal is effected with a single pole changeover switch.

The alternative is to use a permanent magnet motor and reverse the supply polarity. Now only two wires feed the motor but supply polarity reversal needs a double pole changeover switch.

A series or shunt wound motor can be adapted to use this polarity change method of reversal by the simple expedient of feeding the field winding via a bridge rectifier (figure 4-6). Because of the rectifier current always flows in the same direction through the field the field direction is constant and the motor reverses whenever the supply polarity is changed.

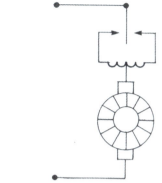

FIGURE 4.5   SPLIT SERIES MOTOR

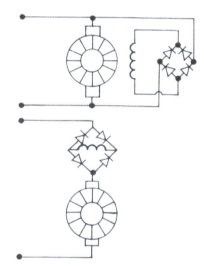

FIGURE 4.6   POLARITY CHANGE REVERSAL - WOUND FIELD MOTORS

4.7 Motor Ratings and Speed Control

The induction motors discussed in earlier chapters have clearly defined operating voltages and shaft speeds determined by the motor windings and the supply frequency. They will only perform satisfactorily if operated close to these rated voltages and speeds.

Commutator motors are much more flexible and their operating characteristics can be varied over a wide range by suitable choice of supply voltage and field strength.

Larger motors commonly carry nameplate ratings for supply voltage, current, shaft speed and rated power. The most important of these is the current rating because it is this that mainly determines the power dissipated within the motor. Provided, apart from temporary overloads, this current is not exceeded the combination of supply voltage, shaft speed and power output can be varied over a wide range to meet the requirements of a particular load.

Constant maximum current means that the maximum torque available from a particular motor is fixed and only slightly affected by speed. Speed can be varied over a wide range by control of armature voltage – 5:1 is possible with simple control gear, ratios in excess of 100:1 are possible with more elaborate arrangements. However, because the maximum torque is fixed, shaft horsepower varies in the same ratio. This means that, if the speed of a motor is reduced by a factor of five by control of armature voltage, the shaft horsepower is also reduced by a factor of five.

The importance of this depends on the nature of the load. With a lathe or a drilling machine the torque required is usually greater at low speed than high and the motor size must be large enough to supply the necessary torque at the lowest speed. If it is desired to vary the speed over a wide range it may be necessary to fit a motor two or three times larger than the minimum size needed for single speed operation.

Fan loads are quite different. The torque required to drive a fan changes with the square of the speed so that if the speed is halved the torque required is divided by four (this is true of both centrifugal and propellor type fans and pumps). This means that if, as is usual, the motor size is chosen for the maximum speed case there will be ample torque available for any lower speed.

All the above discussion assumes that the motor is run at or below its rated speed. Although it is not necessarily looked kindly upon by the motor manufacturer, it is usually possible to operate a motor above its rated speed by increasing the armature voltage and achieve a corresponding increase above its rated horsepower. While this is perfectly feasible with some motors it must be approached with considerable caution – if anything goes wrong you will get no sympathy from anyone – least of all the motor manufacturer!

The biggest problem is the increased centrifugal forces on the armature conductors. If they break loose they will jam the armature with fairly spectacular results. If you are tempted to experiment do make extra sure that the circuit is properly protected with a circuit breaker or fuse and that the motor casing is connected to ground. Also ensure that the motor is properly anchored to something solid and that you are not in the line of fire if anything breaks loose. Keep the field voltage constant at its normal value and increase the armature voltage in small steps. Stop if there is any sign of excessive sparking between brushes and commutator. Keep a safety margin between test conditions and normal running - do not run a motor at more than 80% of the speed reached on the bench test.

Although safety precautions must never be neglected, many small and medium sized motors will happily operate at speeds well above their nameplate rating. The motor on my workshop lathe is an ancient 115V ½H.P. 2000 R.P.M. machine which regularly operates over the speed range 400 to 4000 R.P.M. Small low voltage motors will run at higher speeds but don't try this on high voltage high speed motors (e.g. electric drill and vacuum cleaner motors) as these are already operating near their electrical and mechanical limits and are unlikely to survive the experiment.

All the above comments apply to change of armature voltage. The same flexibility does not apply to the field voltage. Although speed can be changed by varying the field strength the useful range is small. If the field strength is reduced the armature speed increases but, because this reduces the torque produced per amp of armature current, the motor efficiency drops rapidly. Significant increase in field strength over the design value is rarely practicable as this is limited both by

magnetic saturation in the iron circuit and overheating of the field windings.

For speed variation by control of armature voltage, the field strength must be kept reasonably constant. Permanent magnet motors are ideal. Shunt wound motors are equally suitable provided the field is separately supplied with its rated voltage. Series wound and universal motors are convenient in that the armature current automatically maintains the right field conditions almost irrespective of armature voltage. However, this type of motor is less satisfactory for controlled speed applications because, as detailed in section 4.3, changes in load result in large changes in shaft speed.

4.8 Motor Construction
Larger motor types are manufactured in housings not very different from those used for induction motors of similar size and it may need close examination to distinguish between the two types.

Fig. 4.7 A selection of small permanent magnet motors

Fig. 4.8 *Series wound universal motors*

Smaller sizes are made in an almost bewildering variety of forms – some for reasons of mechanical convenience and cost, some to achieve special performance characteristics. Figures 4-7 to 4-9 show some of the more commonly encountered types.

The five motors in the foreground of figure 4-7 are all small permanent magnet motors used in toys and small mechanisms. These are very low cost items mainly intended for operation from battery supplies in the range 1.5 to 6V. The left hand motor in the back row is the capstan drive motor from an audio cassette player. These permanent magnet motors are designed for exceptionally smooth, silent operation at constant speed. They normally operate from six or twelve volt supplies and often have a built in mechanical or electronic speed governor.

The centre and right hand motors in the back row are high quality 24 to 28V D.C. permanent magnet motors designed for military and avionic use.

The right hand motor is built in the very popular size 11 military frame size. The size number is the approximate outside diameter in tenths of an inch - other sizes in the range are 08,15,18 and 23. Many types of military electro-mechanical devices are housed in this series and, apart from frame size, are almost identical in appearance. Fortunately they normally carry reasonably helpful identification labels.

Figure 4-8 shows two series wound universal motors used for main drum drive in domestic automatic washing machines. These are powerful high speed machines that normally operate in conjunction with an electronic speed control system. This type of motor is discussed in more detail in section 9.4.2.

Figure 4-9 shows an open frame series wound stator and armature. This type of construction is often used in power tools and similar devices. To reduce cost and size the motor no longer exists as a separate item. The stator and armature are simply built into the main

Fig. 4.9 *Open frame motor components*

framework which carries the rest of the mechanical and electrical components.

Commutator motors are often used as servo motors. These are motors used for remote positioning and control applications. To achieve rapid and precise positioning they need to have a high ratio of output torque to armature inertia.

For general purpose applications they achieve this by using a long, small diameter armature moving in an exceptionally strong permanent magnet field. The long narrow armature has a smaller moment of inertia than a short fat one of the same volume and the strong field gives a large torque per armature ampere-turn. High temperature insulation is used to permit the large armature currents needed to develop high torques in a small armature. The motor in figure 4-10 is of this type. This particular motor has a

Fig. 4.10 *This servo motor has a tacho-generator fitted (the small end cylinder).*

79

Fig. 4.11 *A disc-armature servo motor disassembled*

small tacho-generator mounted on the non-drive end. This produces a voltage proportional to speed which is used by the control circuits to improve the overall accuracy of response to control signals.

For many applications this type of servo motor is a very good compromise between performance, cost and size. However, for the most demanding applications the presence of iron in the armature limits the maximum achievable torque to inertia ratio. With a conventional copper and iron armature the iron provides a low reluctance path for the magnetic flux and is a convenient mechanical link to transfer the forces exerted on the conductors to the output shaft. The iron itself does not contribute to the output torque and its presence increases the inertia of the rotating element. Because the teeth in the armature saturate at high armature currents, the iron also limits the peak

torque that can be achieved. Fortunately it is possible to completely eliminate the use of iron in the rotor by arranging the armature windings to take the form of either a thin disc or a hollow cylinder.

The disc armature motor is often called a printed circuit motor because the earliest armature construction used photo-etched conductors on an insulating disc - the same techniques as used in electronic printed circuits. However, in order to permit heavier section conductors, most modern disc armature motors use stamped copper segments, welded turn to turn at the periphery to form a continuous pattern of conductors.

A disassembled disc armature motor is shown in figure 4-11. Because there is no iron in the armature to reduce the reluctance of the path followed by the field flux, the total air and copper gap in the magnetic circuit must be kept small. This is achieved by making the total thickness of the disc as small as possible and placing a powerful set of permanent magnets as close as possible each side of the disc.

The arrangement of the conductors is shown in figure 4-12. It is basically a series of current carrying loops formed by radial conductors linked together near the centre and at the periphery. It is only the radial part which generates useful torque, the inner and outer parts simply provide return paths for the current. An extremely neat feature of this form of construction is that the inner return paths fall in exactly the right position to form a commutator with one commutator bar per turn – all that is needed to complete it is two or more brushes mounted axially to bear directly on the conductor pattern at this radius.

In contrast, the outer return paths are all bad news. They generate no useful torque and, because they are at a large radius, they cause a disproportionately large increase in armature inertia (the inertia of a disc is proportional to the *fourth power* of the diameter). To reduce the space taken up by these outer paths it is usual to use a multipolar field - six or eight pole machines are common as this reduces the length of each outer conductor. With a two pole machine each outer conductor would have to span almost 180 degrees. This is reduced to about 45 degrees on an eight pole machine – a fourfold improvement. Although it is possible to interconnect the armature windings so that all four sets of armature turns (two poles per winding so an eight pole machine needs four sets of turns) are driven from a single pair of brushes 45 degrees apart it is usual to fit two pairs of cross connected brushes at 45 degree spacing to simplify the armature and reduce the

Fig. 4.12 *A disc armature*

current density at the brushes.

With conventional armatures there is a small variation in the reluctance of the magnetic circuit as individual armature teeth pass the stator poles which produces an effect known as "cogging". This results in a slight variation in output torque as the armature rotates. If the field only is present, there is a tendency for the armature to settle in one of a number of preferred positions. Many motors use "skewed" rotors – a slight twist is given to the armature laminations when they are assembled on the shaft to smooth out this variation but some cogging always remains.

The ironless rotor machines are completely free from this effect and, because of the very large number of commutator segments (one per "wire") in the disc machines, have exceptionally small variation in torque as the armature rotates. This permits smooth operation down to speeds of only a few revolutions per minute.

Quite apart from its advantages as a low inertia motor the disc armature motor is also useful in applications where length is very restricted and a short, large diameter motor is an advantage. Some electrically driven cooling fans in automobiles are of this type.

It is not easy to make the disc armature motor in very small sizes. Also, in the most demanding servo application, the extra inertia of the outer return paths limits the performance.

An alternative ironless rotor construction, which overcomes the problems, is based on an ironless armature made in the form of a long thin walled hollow cylinder open at one end and connected to a commutator at the other. The armature is made by winding the turns in the right shape on a jig and then bonding the turns together with resin and fibreglass. This is sometimes called a moving coil motor because there are similarities between the construction and that of a moving coil ammeter.

In high power servo motors this hollow cylindrical armature rotates in the annular gap between a central stationary iron cylinder and a powerful external field magnet. In this arrangement it is the long straight axial conductors that produce the torque. The neutral end turns are shorter than in the disc armature and at a smaller radius so that they only slightly increase the armature inertia. Extremely high performance is possible with these low inertia motors. Types used for special magnetic tape drives in computers are capable of two hundred complete start, reposition and stop cycles per second! When this extreme performance is

Fig. 4.13 *An ironless rotor low inertia motor. (Courtesy Portescap (U.K.) Ltd.)*

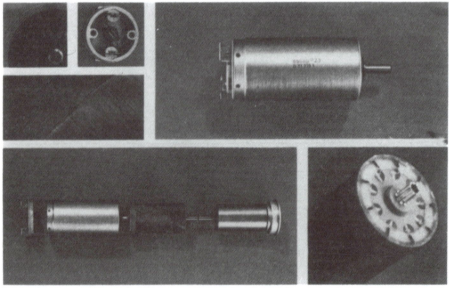

not needed a more compact low inertia construction is possible. In this the external field magnets are replaced by a soft iron tube which forms the motor housing and the field flux is now generated by a small fixed diametrically magnetised cylindrical magnet mounted inside the armature (figure 4-13). These motors can be made in very small sizes and, because they are compact and efficient, are equally suitable for general purpose applications.

In these smaller sizes, instead of forming the armature conductors into a series of straight sided loops, a continuous wave wound construction is used which is, to a large extent, self supporting. An armature of this type is shown in figure 4.13. Each turn of the armature winding forms a loop at approximately 45 degrees to the shaft axis and as the winding progresses the later loops cross over earlier loops at approximately 90 degrees to form a two layer winding of exceptional stiffness and rigidity.

A small diameter commutator is used and this permits the use of precious metal brushes and commutator segments to give exceptionally low contact resistance and reduced friction.

4.9 Electronic commutation

So far all the motors described use mechanical switching to achieve the necessary commutation of magnetic field direction. This is a weak point in the otherwise desirable characteristics of this type of motor. Brushes and commutator need regular maintenance and replacement and these parts wear out long before the remainder of the machine.

Purpose designed integrated circuits now present a solution to this problem. Two general families are available.

The first type essentially duplicates the characteristics of a conventional commutator machine. It requires sensors to be mounted in the motor to monitor rotor position. Magnetic, optical or even mechanical switches could be used for this function but the most convenient and commonly used method is magnetic sensors. These are tiny integrated circuits which are sensitive to the presence of magnetic field and send a signal to the main control circuit whenever a magnet attached to the rotor passes the sensor.

These sensors are pretty sensitive so the magnet need be no larger than a grain of rice!

Although, in principle, any of the standard commutator motor construction types could be controlled in this manner, the universally employed configuration is a permanent magnet rotor with a two or three phase winding on the surrounding stator . The permanent magnet which provides the magnetic field is now located in the rotor and this reversal of function conveniently eliminates the requirement to supply power to the rotating component .

Using the rotor sensor information the main control integrated circuit now switches power to the stator windings in the correct sequence to provide continuous rotary torque. In the simplest case, because the permanent magnet rotor is supplying a constant magnetic field and the full switched supply voltage is applied to the stator windings, the combination behaves in exactly the same manner as a permanent magnet field commutator machine. It is essentially a constant speed machine with the output speed directly proportional to supply voltage and the armature current

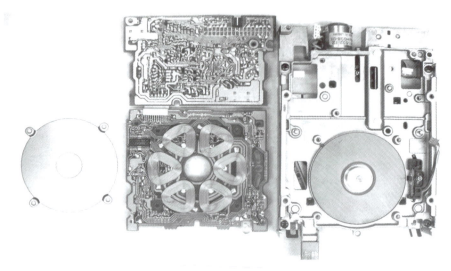

Fig. 4.14 *Floppy disk drive dismantled to show the electronically commutated drive motor*

directly proportional to load torque.

This type of control is mainly used for direct replacement of commutator motors and in cases when control of speed by supply voltage is a desirable attribute. Small cooling fans are a typical application. Some types of controller can be microprocessor controlled and the combination can be programmed to deliver almost any desired speed and load characteristic within the limitations of the frame size of the controlled motor. However, for the more demanding applications the second type of controller is the preferred choice

This is basically similar to the Variable Frequency Drive system used to drive induction motors already described in Chapter 3. However it is also used in somewhat simplified form to control special purpose motor types in a wide variety of applications. In this case the controller once again drives a permanent magnet rotor

working in conjunction with a two or three phase wound stator. The controller includes a fixed or variable supply frequency generator and the programmed switching of power to the stator winding generates a magnetic field rotating at this supply frequency. Because the rotor is a permanent magnet it locks directly to the rotating magnetic field. This is synchronous rotation - there is no "slip" as there would be with an induction motor and precise control of speed is possible.

While both control systems work fine with conventionally constructed motors, the very short optimum magnet lengths of modern ferrite and rare earth supermagnets make radically different configurations possible. Some are essentially of flat thin planar rather than co-axial construction. This results in a much shorter total magnetic path length. Because of this, it is no longer essential to use a laminated ferromagnetic

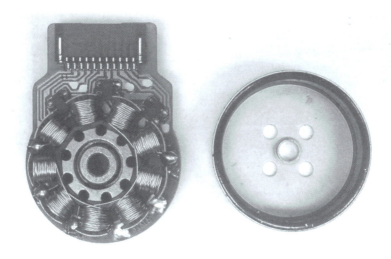

Fig. 4.15 *CD ROM/DVD drive motor, rotor on the right*

return path for the time varying flux. This makes it possible to install the stator windings as a series of thin flat aircored coils mounted on any convenient substrate and this is often the printed circuit board which carries the associated control circuitry. The floppy disk drive motor of figure 4.14 is of this type. The air cored drive coils that are mounted on the central printed board form a set of three phase drive windings arranged as three diametrically opposite pairs. The rotor is a ferrite disk (shown on the right of the figure) that is magnetised with eight alternating N and S magnetic poles. This field passes axially through the coils and the magnetic path is closed through an iron disk (shown on the left) mounted on the other side of the printed board.

With conventional construction, best efficiency and power to weight ratio favour the use of a small number of field poles. Two or four pole machines are the usual

choice. With planar construction where it is important to achieve a comparatively short magnetic path length this is no longer true and multipole machines are common. It is particularly easy to do this when, as in the floppy disk motor the rotor is in the form of thin disk magnetised with an alternating pattern of N and S segments.

With rare earth supermagnets the even shorter optimum magnet length makes possible the radically different design shown in figure 4.15. This is the drive motor from a CD ROM/DVD drive. The fixed portion is constructed in a similar way to the armature of a conventional commutator motor but it is wound as a three pole three phase stator. The rotor is the thin walled black cylinder shown disassembled on the right and is radially magnetised with twelve N S poles . It is mounted within an iron cup which provides the magnetic return path for the radial poles.

Fig 4.16 *Hard disk drive internals*

A rather different category is the moving coil motors used for very high performance servo positioning the magnetic read/write heads of hard disk drives. These only move over a small angular range so that control can be by variable direct current drive with no requirement for mechanical or electronic commutation. A typical disassembled drive is shown in figure 4.16. The disks themselves are driven by a motor of the type described in the previous paragraph. The moving coil read/write head drive is the assembly in the upper left of the figure. The fixed magnetic field is provided by a pair of rare earth magnets. Each magnet is magnetised in the thickness direction with the exposed North and South poles located in two radial sectors separated by a narrow unmagnetised region. The back of each magnet is mounted on a soft iron plate which closes the magnetic circuit by magnetically connecting the back North and South poles. The assembled pair of magnets provides adjacent NS and SN field sectors which are traversed by the current carrying wires in the radial parts of the drive coil.

In the figure the upper magnet assembly has been removed to expose the drive coil which generates the necessary rotary torque needed to rapidly and precisely position the long reading head arm over the correct location on the disk. These motors are capable of positioning the heads to within an accuracy of a few microns in a few thousandths of a second!

CHAPTER 5

Stepper Motors

5.1 General

The motors so far discussed have all been primarily intended for producing power by continuous rotation of an output shaft. Although stepper motors can be used for this purpose, their main use is precise positioning of output shafts in controlled increments. They are widely used in industrial and military control systems and are now being used extensively in consumer electronic products controlled by computers or microprocessors. An electronic typewriter is a typical example. In these, one stepper motor will control the daisy wheel print head to select the character to be printed, a second stepper motor will traverse the printhead along the line and a third will advance the paper feed.

Stepper motors are used in these applications because of their unique ability to move an output shaft to an accurately known position simply by driving the stepper motor with a predetermined number of stepping pulses. This is very much simpler and more adaptable to computer control than the alternative of using a conventional motor which also needs a device to measure shaft position and a control amplifier to give the same

degree of control of output shaft position.

Stepper motors can be broadly classified into permanent magnet and variable reluctance types and each type can be wide angle stepping (4-24 steps per revolution) or narrow angle stepping (50-200 steps per revolution). Maximum stepping rates for high performance narrow angle stepping motors can exceed 5000 steps per second (i.e. 3,000 R.P.M. on a 100 step/rev machine) but stepping rates of a few hundred pulses per second are more typical, particularly in wide stepping angle designs. This means that stepper motors are mainly low speed machines with modest power outputs for a given frame size. However, this is a small price to pay for the extreme convenience of controlled step operation.

5.2 Stepper Motor Operation

The following discussion mainly refers to permanent magnet stepper motors as these are the most widely used variety. Reluctance motors are covered in paragraph 5.3

Permanent magnet stepper motors are really a variety of synchronous motor – in fact a permanent magnet

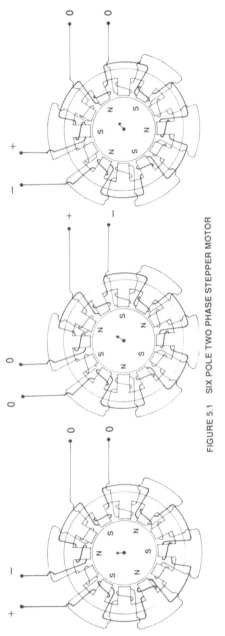

FIGURE 5.1 SIX POLE TWO PHASE STEPPER MOTOR

synchronous motor can be used as a stepper motor and vice versa. The difference is in how the motor design is optimised for the two different types of application.

Synchronous motors are mostly optimised for maximum power output from a given frame size. Except in the very small sizes, they will overheat if the rotor is prevented from rotating.

Stepper motors are designed to operate continuously with the rotor stationary and generate a strong restoring force (i.e. holding torque) if the rotor is moved away from its rest position. Accuracy of stepping position and high maximum stepping rate are important features in stepper motors. Because in a single step the motor has to accelerate up to what is, in effect, synchronous speed and then come to a halt, a high torque to inertia ratio is needed if high stepping rates are to be possible. Most characteristics improve as the number of steps per revolution increase and stepper motors are made with as many as 200 steps/rev. This large number of poles means that, for a given frame size, the power output is rather small. To obtain a better power to size ratio fewer steps/rev are used. For precision steppers 50,60,100 and 200 steps/rev are popular. When power output is more important wider step angles are used – typically 6,8,12 or 24 steps/rev.

Most permanent magnet stepper motors use a multipole permanent magnet rotor rotating inside a two phase stator winding. A two phase stator winding is used as this is the minimum number of phases that will sequentially step the rotor in a given direction. A six pole (12 step) two phase stator winding is shown in figure 5-1 in three successive step positions. In each case the rotor comes to rest with opposite poles adjacent, rotor to stator, as this is the zero torque position with the magnetic forces balanced. As the rotor is displaced from this position a restoring torque is generated which increases approximately sinusoidally (figure 5-2) to a maximum at a displacement of one stator tooth (i.e. one complete step away) with the rotor magnetic poles then half way between the energised stator teeth. If the rotor is displaced further in the same direction the restoring force drops and then changes sign. If it is released at this point it will move to the next stable position four steps away from the original position. The peak value of this torque curve is the maximum torque load that can be placed on the motor, at rated current and zero speed, without it slipping out of step and is known as the holding torque. In practice a safety margin is needed and about 70% of this value is as much as can be safely used. At this value the rotor will lag the ideal position by about half a step.

If the windings are not energised the permanent magnet rotor will move to the nearest low reluctance position with the rotor magnetic poles opposite stator teeth. Because there is one stator tooth for each step, these detent positions occur four times as often as the main holding torque positions (see figure 5-2). The amount of the detent torque is less than the main holding torque and varies greatly with motor design. If required, it can be reduced to almost nothing by using stator laminations with very small gaps between the tips of the teeth so that the magnetic reluctance seen by the rotor is almost constant.

Stepper motors can be used in applications that need most of the torque that they can generate. In these cases the error in shaft position can

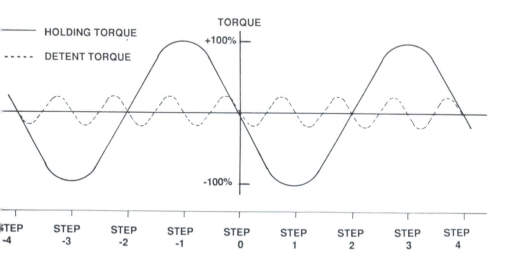

FIGURE 5.2   HOLDING TORQUE AND DETENT TORQUE

approach half a step as this is where the maximum usable torque is generated. They can also be used in applications where the load is very light but high positional accuracy is essential. In this case the pattern of errors is different. Errors arising from the mechanical construction of the stator and rotor can be very small because of the inherent symmetry of construction and the fact that the final field direction is the magnetic average of the flux from all of the many pole pieces. Some errors arise if there are small differences in the windings, or in the currents which are applied to the windings in the stepping sequence, but these are quite small – typically less than 5% of one step (on a 100 step motor this is less than $\frac{1}{5}$ degree). Not only is this error non-cumulative (i.e. the error on one step does not add to the error on the next or subsequent steps) but it cancels out to zero every fourth step. The reason for this is that, with a two phase stepper, there are only four different current patterns applied to the windings and these repeat every four steps. This can be seen in figure 5-3 where the excitation patterns repeat every fourth step. To take advantage of this, in precision applications, it is common to use stepper motors in the "4-step mode" in which all positions are multiples of four steps.

It is also possible to operate stepper motors in a fractional step mode. In normal operation each winding is supplied with full current in sequence. In fractional stepping, a full current single winding step is followed by a step in which the current is divided between the

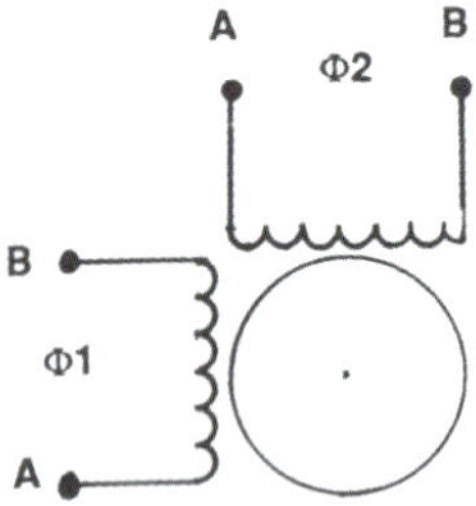

		Φ1		Φ2	
		A	B	A	B
SINGLE	1	+	-	0	0
COIL MODE	2	0	0	+	-
	3	-	+	0	0
	4	0	0	-	+
	(5)	+	-	0	0
DOUBLE	1	+	-	+	-
COIL MODE	2	-	+	+	-
	3	-	+	-	+
	4	+	-	-	+
	(5)	+	-	+	-
HALF	1	+	-	+	-
STEP	2	+	-	0	0
MODE	3	-	+	+	-
	4	0	0	+	-
	5	-	+	-	+
	6	-	+	0	0
	7	+	-	-	+
	8	0	0	-	+
	(9)	+	-	+	-

FOR REVERSE SEQUENCE
INTERCHANGE A & B COLUMNS
ON Φ1 OR Φ2

FIGURE 5.3   TWO PHASE TWO COIL
STEPPER MOTOR

90

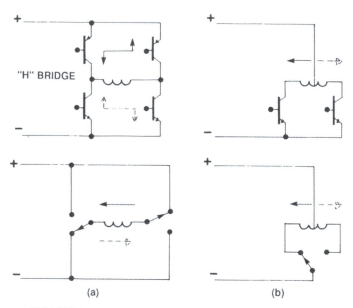

"H" BRIDGE

(a)   (b)

FIGURE 5.4   TRANSISTOR AND MECHANICAL SWITCHING
ARRANGEMENTS

first winding and the next winding in the stepping sequence. Any intermediate rotor position can be achieved by proper choice of the ratio between the two currents but it is normal to choose a simple 1:1 ratio which results in a single half step position, thus doubling the number of steps. This is known as the "half step mode".

Two phase stepper motors are mostly low voltage (6 to 24 v nominal) machines designed to be driven by transistors or one of the special integrated circuits which are designed for this purpose. The simplest type is the two coil motor shown in figure 5-3. It can be operated by sequentially energising single coils as shown in table A or by energising coils in pairs as shown in table B. Table B is normally used as it is slightly more efficient than table A because it makes better use of the

windings. In A the rest position of the rotor poles is immediately adjacent to the energised stator poles, in B the rest position is half way between the pairs of energised poles, half a step different from A. Table C alternates between the single coil and double coil mode and results in the half step mode mentioned earlier. Only four and eight steps are shown in the tables as in each case all subsequent steps are repeats of the initial series.

The above motor is often called a bipolar stepper motor because it is necessary for both forward and reverse current to flow through each coil in the course of the stepping sequence. This complicates the switching arrangements and if transistor switching is used it needs four transistors in an "H" bridge arrangement for each coil (figure 5-4a). To simplify the switching, four coil steppers

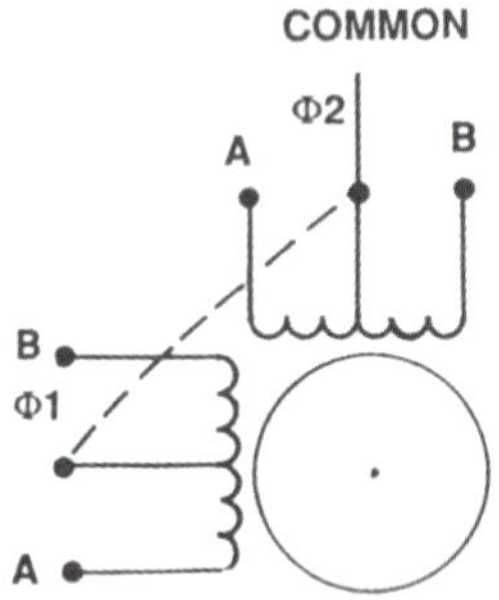

		Φ1		Φ2	
		A	B	A	B
A SINGLE COIL MODE	1	ON	-	-	-
	2	-	-	ON	-
	3	-	ON	-	-
	4	-	-	-	ON
	(5)	ON	-	-	-
B DOUBLE COIL MODE	1	ON	-	-	ON
	2	ON	-	ON	-
	3	-	ON	ON	-
	4	-	ON	-	ON
	(5)	ON	-	-	ON
C HALF STEP MODE	1	ON	-	-	ON
	2	ON	-	-	-
	3	ON	-	ON	-
	4	-	-	ON	-
	5	-	ON	ON	-
	6	-	ON	-	-
	7	-	ON	-	ON
	8	-	-	-	ON
	(9)	ON	-	-	ON

FIGURE 5.5 TWO PHASE/FOUR PHASE FOUR COIL STEPPER MOTOR

are available. Each of the windings is split into two coils which permits current reversal to be obtained with only two switches (figure 5-4b). This is much more convenient and most small two phase steppers are of this type.

These steppers are referred to as four phase or unipolar two phase steppers. The switching sequences are shown in figure 5-5.

5.3 Stepper Motor Control Systems

Although the switching needed to drive stepper motors can be carried out by rotary mechanical switches or commutators (the old World War II "M" motors were driven this way) special purpose integrated circuits are now almost universally used for this purpose, either directly or, for the larger motors, via power transistors. These devices greatly simplify matters by automatically applying the right sequence of currents to the stepper coils in response to a simple train of pulses at the input pin – one pulse for each output step. Forward or reverse rotation is selected by switching the voltage applied to a second input pin.

Some of the devices available for this purpose are shown in the following table.

Device Type	Manufacturer	Remarks
SAA1027	Signetics/ Mullard	350mA 4phase step generator/driver
SAA1042	Motorola	500mA 4phase step generator/driver with full and half step modes
UCN-4204B	Sprague	1.25A 4phase step generator/driver with full and half step modes
L293D	SGS	2 phase step generator/driver

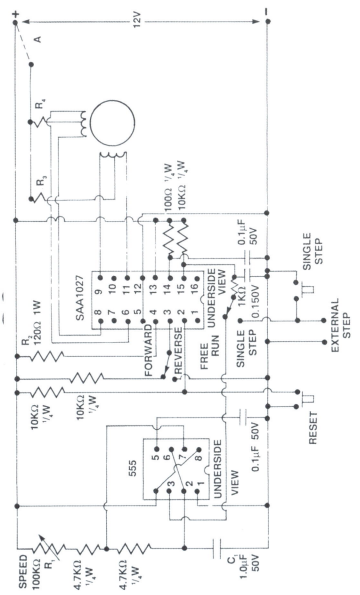

FIGURE 5.6   STEPPER MOTOR DRIVE CIRCUIT

93

TEA1012	Signetics/ Mullard	50mA 4 phase step generator with full and half step mode and current control
L297	SGS-ATES	2/4 phase step generator with full and half step mode and current control
L298	SGS-ATES	H bridge power stage for L297
UDN-2878/9	Sprague	Quadruple 4A power driver for 4 phase steppers
555	Motorola/ National Signetics/ Texas	Industry standard timer/pulse generator

Each of the first three devices is a complete stepper motor drive system which generates the correct sequence of drive pulses at a power level high enough to drive small motors directly. The SAA1027 is a popular basic device for driving motors in the full step mode. The SAA1042 provides slightly more output current and can be operated in both full step and half step mode. The UCN-4204B is a higher power 15 v 1.25A device and is also available in a 35 v version (UCN-4205-B)

The TEA1012 and L297 are more complex devices with limited output drive capability (typically 50mA) intended to be used with an array of four power transistors to drive a 4 phase motor or with an "H" bridge power stage for a two phase motor. In addition to both full step and half step operation they also contain special circuitry to control the current taken by the motor.

The performance of a stepper motor at high stepping rates is degraded because the stepping voltages are not applied to the windings for long enough during each step for the current in the winding to build up to its full steady state value. The rate of build-up can be greatly improved by increasing the voltage applied to the windings to a value well above the maximum continuous stationary rotor rating with a corresponding improvement in high speed torque. However, this would lead to excessive current and overheating at low speeds. The TEA1012 and L297 both include "chopper" circuits which interrupt the current supply to the motor at a supersonic frequency. The ratio of the "on" to "off" time of this chopped current is automatically controlled to maintain the average value constant over the useful speed range of the motor. This is known as operating the motor in a constant current mode and is one of the best ways of extracting maximum performance from a particular stepper motor.

The L297 and UDN-2878/9 are power stages that can be driven from a lower power step generator or directly from a suitably programmed microcomputer.

The last device is a popular wide range pulse generator (less than one pulse per second to hundreds of thousands of pulses per second!). It is used to provide the pulse input to the stepper motor drivers as a convenient method of controlling motor speed.

A typical arrangement for a simple stepper motor driver is shown in figure 5-6. This is capable of driving 12V 4 phase (i.e. unipolar) stepper motors at up to 350 mA per phase.

On free run the stepping speed is controlled by R1 and C1. For the values shown the range is approximately 10 to 100 steps/sec. Higher speeds can be obtained by reducing the value of C1. Single stepping is possible either by the push button or by an external contact.

When first switched on the SAA1027 may start at random at any step in its repeating four step sequence. If the start state is important, grounding pin 2 via the RESET button overrides all other inputs and sets the SAA1027 to step 0.

R2 is the bias resistor which sets the current capability of the output stages. With this set at 120 ohms the SAA1027 will drive any motor at up to 350mA per output (this corresponds to approx. 30+30 ohms winding resistance per phase for a 12V motor). If very small motors are used consuming perhaps only 50mA per output it is permissible to increase the value of R2 to reduce the power wasted in it. However, if R2 is increased, to ensure proper switching of the output stage, the current through it should always be at least 25% of the required output current.

R3 and R4 depend on the motor fitted and limit the maximum current through the motor. If the current taken is less than 350mA per output then both R3 and R4 can be zero. If a low voltage motor is used (some motors are rated at 5V or less for continous operation), or a high power motor is used that could overload the SAA1027, then R3 and R4 are chosen to limit the motor voltage and current to a safe value. Besides limiting the motor dissipation, as explained later, these resistors perform the useful function of "current forcing" which improves the performance at high stepping rates.

If a very low voltage motor is used it may be preferable to break the connection marked A in figure 5-6 and feed the motor from a separate low voltage supply. The SAA1027 still needs its own nominal 12v supply as it must have a supply that is always higher than 9.5v if it is to function correctly. If this is a problem the SAA1042 can be used as this can operate from a supply as low as 5v.

If the power output of the SAA1027 is insufficient it can be boosted with a power amplifier — Figure 5-7 shows a suitable arrangement. This can run from the same 12v supply as the SAA1027 driver or from a separate supply anywhere in the range 6 to 18v. With the values shown this boosts the output from 350mA to 5A. If this is more than necessary it can be reduced to the required value by choice of the value of R10 and R11 and/or the supply voltage in the same way as R3 and R4 in figure 5-6. The type of transistor used is not at all critical and, provided its current rating is not exceeded, almost any audio P.N.P. power transistor can be used.

Diodes D1 to D9 are there to protect the power transistors from high voltage switching transients caused by the rapid changes of current in the stray inductance of the windings. The size of these transients is very dependent on the detail design of the motor. In many cases, particularly at low supply voltages, these diodes can be safely omitted. However, they are very low cost components and it is better to be safe than sorry!

5.4 Operation at High Stepping Rates
Most of the discussion so far has been on the performance of stepper motors at low and medium stepping rates. At these rates the motor starts and stops in response to individual steps. If a burst of steps is applied the motor will move the required number of steps without missing any at the beginning or overshooting at the end. This is only possible at moderate stepping rates and with low inertia loads.

The first problem is that at high stepping rates there is not sufficient time for the current to build up to its full value

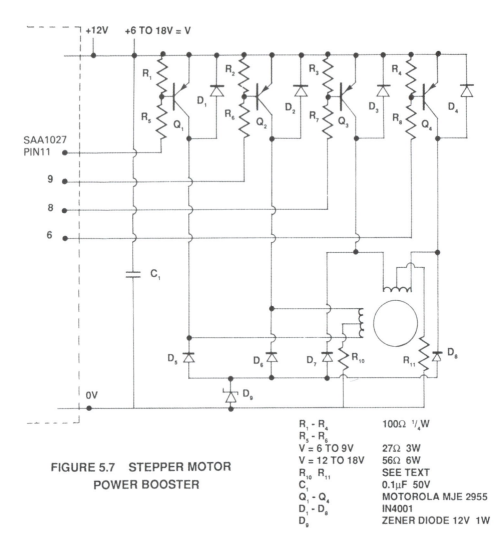

$R_1 - R_4$	100Ω $\frac{1}{4}$W
$R_5 - R_6$	
V = 6 TO 9V	27Ω 3W
V = 12 TO 18V	56Ω 6W
R_{10} R_{11}	SEE TEXT
C_1	0.1μF 50V
$Q_1 - Q_4$	MOTOROLA MJE 2955
$D_1 - D_8$	IN4001
D_9	ZENER DIODE 12V 1W

FIGURE 5.7   STEPPER MOTOR
POWER BOOSTER

in the duration of one step. In addition to this, as the motor gains speed, the permanent magnet rotor induces a back E.M.F. in the stator windings which opposes the applied voltage and further reduces the current in the windings.

The constant current mode of the TEA1012 and the L297 is a very elegant solution to both these problems. A somewhat simpler brute force solution is to operate a low voltage motor from a high voltage supply but limit the maximum current in the windings to a safe value by external series resistance e.g. R3 and R4 in figure 5-6. This is known as "current forcing". Because the rate of build-up of current in the windings is controlled by the L/R inductance to resistance ratio of the windings, an added external resistance about equal in value to the motor winding resistance will double the rate of build-up. It will also double the resistive losses but that may well be an acceptable penalty.

The second problem is that, when starting up, the available torque may be insufficient to accelerate the rotor plus load inertia to move from standstill through one step interval in the time occupied by two successive stepping pulses. A similar problem is encountered in preventing overshoot when suddenly stopping a motor that is stepping at high rates. If the rotor position lag or overshoot reaches two step positions the motor will slip steps or, in adverse cases on start up, will fail to move at all. This is a fundamental problem and the only cure is to avoid violent changes into, or out of, high stepping rates. Under computer control it is fairly easy to circumvent this problem by arranging the program so

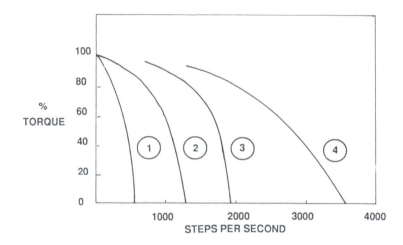

100 STEP/REV MOTOR ON FRICTION LOAD

(1)   PULL-IN TORQUE AT RATED VOLTAGE AND CURRENT

(2)   PULL-IN TORQUE WITH CURRENT FORCING RESISTOR

(3)   SLEWING PERFORMANCE WITH CURRENT FORCING RESISTOR

(4)   SLEWING PERFORMANCE WITH CONSTANT CURRENT DRIVE

FIGURE 5.8   PULL-IN AND PULL-OUT TORQUE CURVES

that the first and last few steps in each burst of steps occur at a lower rate within the acceleration/braking capability of the motor.

Stepping motors vary widely in design and performance and the performance of any individual motor is strongly affected by the drive arrangements. However, figure 5-8 gives a general picture of the sort of performance to be expected from a 100 steps/rev motor in its principal modes.

Curve one shows the pull-in torque available when the motor is operating into a friction load at its continuous rated voltage and current. The torque available falls off fairly rapidly as stepping rate increases because of the effect of winding inductance and the back E.M.F. The motor will operate satisfactorily in an instant start-stop mode anywhere in the region to the left of this curve.

Curve two shows the improvement in performance when the supply voltage is doubled and the current limited to the rated value by a current forcing resistor equal in value to the winding resistance.

Curve three shows the pull-*out* torque with the same supply voltage and current forcing resistor. Operation in this region to the right of curve two is only possible if the motor is first ramped part the way up its speed curve by a few steps at a lower rate. A similar ramp is needed on stopping if it is to stop without overshoot. This is known as the slewing performance of the motor. In this region between curves two and three the speed of the motor can be controlled but instant ("instant" means reaching demanded stepping rate or coming to a halt within one stepping pulse interval) start/stop operation is not possible.

Curve four is an example of the extension in slewing performance possible with a constant current drive system. With high supply voltages and appropriate drivers speeds as high as 10,000 R.P.M. are possible. However, it is comparatively rare for speeds higher than one tenth of this to be used as, in high speed positioning applications, much of the time is spent ramping the motor up to speed and down to a halt and the comparatively small amount of time saved by a very high maximum speed is not worth the additional complication.

All the above comments refer to the motor driving a load in which friction forces dominate. If the motor is driving a high inertia load such as a flywheel, or a load with very low friction, then inertia forces dominate. This gives rise to two problems.

Firstly the inertial forces subtract from the torque available at start up and add to the braking forces needed in the stop mode. This is a nuisance but not too bothersome as it simply dictates the use of a motor with an adequate margin of torque capability.

The second problem is potentially more serious. The interaction between the torque compliance of the motor (i.e. the springiness of the restoring force which centres the rotor on its correct position) and the total inertia of the rotor and load makes the rotor behave as a weight suspended on a spring which oscillates many times back and forth before finally coming to rest. In extreme cases, when the final step is applied, the oscillation amplitude can exceed the maximum permissible dynamic error of two steps (see figure 5-2). At this point the restoring force changes sign and, in successive oscillations, the rotor may

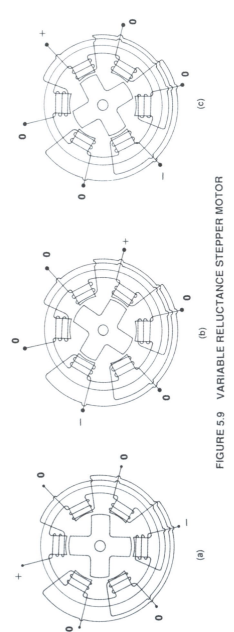

FIGURE 5.9 VARIABLE RELUCTANCE STEPPER MOTOR

then slip an unpredictable number of steps in either direction.

This can be overcome by electrical and/or mechanical damping. Friction losses within the motor, together with eddy current and hysteresis losses in the iron, all tend to damp out the oscillation and, in many cases, this is sufficient. However, in some cases external damping is necessary.

Several methods are possible, either alone or in combination. Provided there is sufficient spare torque, the simplest and most direct method is to apply sufficient friction to the output shaft. This can be either "Coulomb" friction in the form of a dry contact brake or viscous friction in which the surfaces are separated by oil or grease of suitable viscosity. Coulomb friction produces a damping force that is only slightly dependent on speed and in fact rises somewhat at very low speeds – the stick/slip region. This means that the maximum value of the damping force is present as the motor comes to a halt. This is very good for damping but, unless the motor has an ample torque margin, the high static friction can result in a positional error of a large fraction of a step. Dependent on the characteristics of the load and the speed of the motor as it approaches the final step, the rotor may or may not overshoot the correct rest position. This means that the friction error is not a simple lag behind the ideal position – if the rotor finally comes to rest after an overshoot it will lead the ideal position by a roughly equal amount, doubling the total error band.

Provided there is no metal to metal contact the viscous friction system provides a damping force which is directly proportional to speed. This

99

means that as the rotor approaches its final rest position the damping force falls to zero and the positional accuracy of the motor is not degraded. Unfortunately the corollary to this is that the damping force rises with speed and this may severely limit the maximum stepping rate. The optimum solution depends on the application and in some cases a combination of both is appropriate.

External damping can also be provided by the drive circuit arrangements. If sequential single coil driving is used (figure 5-3) and the coil not currently energised is short circuited, eddy currents in the short circuited coil provide strong damping forces. The same effect can be produced in four phase motors by short circuiting the un-energised winding segments. While these are attractive systems their use is limited by the fact that they severely complicate the drive switching arrangements.

In computer-driven stepper motor systems there is much scope for low cunning in the control program to circumvent the problem. The optimum velocity profile can be calculated in advance and stepping rates controlled to give carefully tailored acceleration and deceleration curves. The stepping rate can be controlled to accelerate the motor continuously in the first half of the positioning cycle followed by continuous deceleration to the final rest position to achieve the lowest overall positioning time.

5.5 Variable Reluctance Stepper Motors
Although permanent magnet (P.M.) stepper motors are the most commonly used types, variable reluctance (V.R.) types are often encountered. These motors work on the principle of magnetic attraction acting on projections on a soft iron rotor. They have the advantage that they are easily made in small sizes and with large numbers of steps per revolution.

The principle of operation is shown in figure 5-9. The rotor is made of easily magnetisable soft iron laminations with four projecting teeth. Unlike P.M. steppers, the soft iron rotor cannot tell the difference between a North pole and a South pole and so two phases are not enough. To be directly equivalent the reluctance motor would need four phases but in fact three is sufficient and this is the number that is normally used. In figure 5-9 the three opposite pairs of poles are wound to form the three phases. When any phase is energised the rotor is attracted to the minimum reluctance position which is when a pair of rotor teeth line up with the pole pieces of the energised phase. In figure 5-9a the rotor is lined up with phase 1. Figure 5-9b and Figure 5-9c show that, as each successive phase is energised, the rotor moves one third pole pitch, moving through one complete stator pole pitch each time the cycle of three steps is repeated. In this case 18 steps form a complete revolution. This is a much smaller stepping angle than the nearest equivalent P.M. stepper. A P.M. stepper with six wound pole pieces would have only six steps per revolution.

V.R. stepper motors can be used in much the same way as P.M. steppers. Integrated circuit step generator/drivers are not readily available for them but the necessary drive sequences can be produced using standard reversible counters hard wired into a ring of three sequence. However, unless you are an electronics buff P.M. steppers are easier.

Fig. 5.10 *Four small stepper motors*

One difference that may or may not be an advantage is that V.R. steppers have no significant detent torque. When power is removed the magnetism retained by the soft iron rotor is negligible and the rotor spins freely.

5.6 Stepper Motor Construction

The main problem in stepper motor

Fig. 5.11 *An annular stator*

design is how to cram a large number of steps per revolution into a limited frame size without making too many sacrifices in efficiency and power rating. Much ingenuity has been expended on this and a wide range of different designs are available. Some of these are shown in figure 5-10.

Reading from left to right the first stepper is a double rotor/double stator P.M. stepper. In a conventional P.M. stepper it is difficult to pack a large number of poles into a single stator because of the necessity of providing a phase one winding followed by a phase two winding on successive teeth. If a large number of poles is needed this leads to a very complicated series of windings. The double rotor/double stator design sidesteps this problem by linking two identical rotor/stator pairs on a single shaft.

All phase one poles are on the first rotor stator pair. The second pair is identical but the rotor is rotated one step (90 electrical degrees) to provide all the phase two poles. If conventional stator laminations are used this allows at least double the number of poles to be accommodated. This design takes it still further by using the annular stator

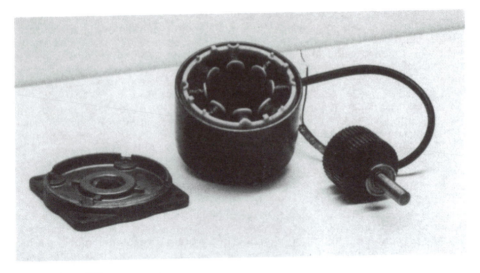

Fig. 5.12 *SLO-SYN stepper motor*

design shown in figure 5-11. This is similar to the design used in synchronous motors for some types of electric clocks. A single large diameter coil is used concentric with the motor shaft. This is surrounded by a soft iron housing which, on its inner surface, is broken into teeth projecting inwards from alternate ends. A hundred step/rev. motor is easily produced using only two coils, one in each stator. Each of the two stators has 2x25 teeth pointing in alternate directions providing the fifty poles necessary for each phase.

The main limitation of this design is that it is not possible to laminate the stator iron circuit and because of this the eddy current losses are high. This results in the efficiency falling off rather rapidly at stepping rates higher than a few hundred pulses per second.

A similar construction system can be used for V.R. steppers but in this case three rotor/stator pairs are needed to provide the three phase sequence.

The second motor in figure 5-10 is a World War II vintage "M" type repeater motor. This is a straightforward four pole three phase design with a cylindrical four pole permanent magnet rotor giving six steps/rev. Although this is a P.M. stepper it is designed to have almost no detent torque. This allows the rotor to be driven to any desired intermediate position by choosing the ratio of the currents supplied to two successive phases.

The third motor in figure 5-10 is a modern SLO-SYN 200 steps/rev. P.M. stepper (SLO-SYN is a trade name, of course) and is shown disassembled in figure 5-12. This uses a different method of packing the equivalent of a 100 pole two phase stator in a motor only 3 in. diameter.

The rotor is a powerful, axially magnetised permanent magnet fitted with two soft iron end caps each having fifty teeth on its outer circumference. All the teeth on the first cap will be North poles and all on the second South poles. The teeth in the two caps do not line up – the second cap is rotated half a tooth pitch with respect to the first. Because the pack of laminations that forms the stator extends over the full length of both end caps, as far as the stator is concerned, this is the equivalent of a hundred pole permanent magnet rotor with alternate North and South poles round its circumference.

A straightforward 200 step/rev stepper would need a hundred pole two phase stator winding. This would need 200 pole pieces in its stator – 100 for phase one and 100 for phase two. However, provided the pole pieces actually present are in the right relative position, not all of the 2x100 pole pieces need be present. This stator takes advantage of this by providing only forty pole pieces distributed in eight bunches of five teeth round the inner surface of the stator. Each bunch is surrounded by a single winding. Each alternate bunch is connected to phase 1 and the remaining alternate bunches to phase 2.

The stator teeth are cut on a 48 teeth/rev pitch and this, in conjunction with the 50 tooth rotor, gives a vernier effect so that there is only close alignment between pairs of rotor and stator teeth at opposite ends of a single diameter. The vernier action is such that, when the rotor moves through one step, i.e. 1/200 turn, the diameter at which closest alignment occurs moves 1/8 turn i.e. 50/50−48 = 25 times further. This is exactly the right amount to line up the next bunch of five teeth on the pole piece excited by the next phase. The result of this is that, if the normal sequence of two phase stepping pulses is applied to the phase 1 and phase 2 windings, the rotor will move in 1.8 degree increments, i.e. 200 steps/rev.

This design makes efficient use of the permanent magnet material in the rotor and also uses a fully laminated stator construction. This permits successful operation at the high stepping rates necessary for the more demanding applications. Typical applications are printhead and paper feed control on high speed printers or precise positioning of small mechanisms in machine tools and similar devices.

The last item in figure 5-10 is a fairly conventional 24 steps/rev three phase V.R. stepper. It is included to show the friction brake which is fitted to the shaft extension to provide damping when driving high inertia loads. The brake is a simple brass disc rotating with the rotor shaft and spring loaded into contact with a stationary nylon friction pad.

Figure 5-13 shows a very high performance modern stepper made by the Swiss company Portescap which takes full advantage of the improvement in performance made possible by the use of samarium/cobalt high energy permanent magnet materials. These permit much smaller volumes of magnetic material to be used in the rotor while still maintaining or increasing the maximum torque capability. These alloys have a very high coercivity – this is the characteristic which controls the minimum length of a magnet needed for a particular application. The length of the magnet needed is so short that it possible to make a very low inertia rotor in the form of an axially magnetised disc only 0.028in./0.7mm thick. This also makes it possible to magnetise the rotor with very closely spaced alternate N and S poles. The rotor in figure 5-13 is magnetised in the direction of its thickness into 50 adjacent magnets so that each side has a circle of 50 alternate N and S poles on its surface.

Fig. 5.13 (above)
*An Escap disc rotor
stepper motor*

Fig. 5.14 (right)
*Escap stepper motor
with Velocity Pick-off
Coils.
Both photos courtesy
Portescap (U.K.) Ltd.*

While this is an elegant solution to the manufacture of a fifty pole rotor there remains the problem of producing the multipole two phase stator. A full stator requires 50 + 50 pole pieces and would be excessively complex so a "skeleton" stator system is used. A total of 20 poles is used in two bunches of ten, one bunch for each phase. These are not interwoven but located in separate arcs, each occupying about 180 mechanical degrees.

Each of the poles actually present occupies one of the pole positions of a full 50+ 50 pole stator; the remainder are omitted. This is not a vernier system – all phase 1 poles occur at odd divisions of a 100 division uniformly divided circle and all phase 2 poles occur at even division positions of the same circle.

The construction can be seen in figure 5-14. Each pole consists of two small stacks of "C" shaped laminations, one mounted each side of the rotor disc. The outer ends of the "C" are ground flush so that, when the two halves of the stator assembly are bolted together, these two surfaces butt together and complete the magnetic circuit without significant air gap. The inner ends form the actual pole pieces and are cut back so that, when finally assembled, a small clearance exists between the pole pieces and the active area of the rotor disc. A pair of coils, one each side of the rotor disc, surrounds all the phase 1 poles and can be connected in series or in parallel. A similar pair surrounds the phase 2 poles.

Because of the very low armature inertia, this type of motor can step at very high rates. Start/stop operation is possible with stepping rates as high as 1000 steps/sec. With current forcing slewing rates can reach 20,000 steps/sec.

These motors can be used in the most demanding applications and are often used under direct computer control. The best program strategy for minimum positioning time needs to monitor the rotor speed and direction of rotation at different points in the positioning cycle. A speed and direction readout can be provided by a pair of additional velocity pick-off coils mounted between the two sets of main phase windings. Figure 5-14 shows the location of these coils in the stator assembly.

The rotor induces an A.C. voltage in these windings directly proportional to speed. The pick-offs are mounted in vacant phase 1 and phase 2 positions, 90 electrical degrees apart. This means that the two output waveforms are in quadrature, i.e. one waveform leads or lags the other by 90 degrees. The lead/lag relationship is determined by the direction of rotation and reverses when the rotor reverses. This phase relation is used by the computer to monitor direction of rotation.

CHAPTER 6

Speed Control and Electric Braking

6.1 Speed Control

6.1.1 Induction motors

Induction motors are basically constant speed machines with the shaft speed closely related to the supply frequency. If, as is usually the case, the supply frequency is fixed, opportunities for convenient control of speed are very limited.

One fairly satisfactory system is the use of a double wound stator to provide two fixed speeds. Two sets of windings are inserted into the stator slots, one for each of two speeds. Popular pairs are two pole and four pole or four pole and six pole windings corresponding to 2:1 and 1.5:1 speed change.

A two speed motor is somewhat larger than a single speed motor when both are rated at the higher of the two speeds. This is because of the extra space taken up in the stator by the second set of windings.

The torque available is determined by the frame size and is almost independent of speed. Because of this the horsepower drops at the same rate as the speed – with a 2:1 ratio only half power is available at the lower speed. This is no problem on a fan load, a

centrifugal pump or a mainly viscous friction load as these all need maximum torque at maximum speed and the motor will have power in hand at any lower speed. More care is needed with machine tools. Lower speeds are often needed because larger diameter or tougher workpieces are being machined. The horsepower required will often increase at the lower speed and this must be borne in mind when selecting the size of the motor.

Most induction motors are fitted with low resistance squirrel cage rotors as these permit the motor to operate at high efficiency and with a good constant speed characteristic – only a few per cent change from no load to full load. As discussed in chapter 2 the maximum torque peaks at a speed a little below full load and drops at any lower speed. This is most unsuitable for speed control systems. For the speed to be stable the available torque should increase as the speed decreases in order to restore the speed to the selected value.

Induction motors specially designed for speed control, or for use as servo motors, are fitted with high resistance rotors. This results in a torque characteristic shown in figure 6-1 in

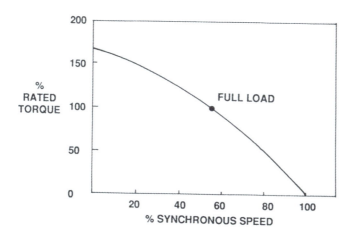

FIGURE 6.1   HIGH RESISTANCE ROTOR
TORQUE/SPEED CURVE

which the torque increases almost linearly as the speed drops. These motors are less efficient than standard types. At full load and rated voltage the shaft speed is only half to two thirds synchronous speed and the power that is equivalent to this loss of speed is dissipated in the high resistance rotor. Because of this the efficiency is usually less than 50% and this limits this technique to small motors delivering only a few watts of mechanical power. It is commonly used in very small motors for military and avionic applications, operating from the 26v or 115v 400Hz high frequency power supplies used in aircraft and some naval installations.

Motors of this type can be controlled over a wide speed range by variation of supply voltage. Single phase motors are not generally suitable for this duty because of the difficulty of switching starting windings in and out of circuit at the right time. Three phase machines can be used but two phase machines are usually more convenient. Small two phase machines have the advantage that one phase can be left permanently connected to the full supply voltage via a capacitor which provides approximately 90 degrees phase shift. Both speed and direction of rotation can then be controlled by the voltage applied to the second phase. This is ideal for small servo positioning systems using synchros or similar transducers which output their error signals as an A.C. signal at supply frequency. This error signal, via a servo amplifier, can power this second phase directly to control the motor speed and direction.

When used in this way it is important that the motor does not continue to run when the second phase voltage is reduced to zero. With a low resistance rotor machine the circulating currents in the rotor are high enough to permit efficient operation with only one phase connected and this is the normal method of operation of single phase motors. However, in the high resistance rotor designs, the rotor currents are very much smaller and the available torque falls rapidly as the second phase voltage is reduced. A completely unloaded motor may just continue to run after the second phase voltage is removed but the available torque is so low that the friction load of a normal gear train is usually sufficient to bring it to a halt.

It must be emphasised that the above comments apply only to small two phase motors designed to run with one phase energised and the rotor stationary. Attempts to run larger, more efficient, motors in this mode would result in severe overheating of the energised phase.

A totally different method of induction motor speed control is by the use of a variable supply frequency. This is an excellent method and is covered in Chapter 3. Although single phase machines can be controlled by this method, starting arrangements are difficult so three phase motors are normally used.

Some of the commonest applications are industrial high speed internal grinding spindles, spindle moulders and routers for wood. These all need speeds in the range 20,000 to 50,000 R.P.M., far beyond the direct drive 3,000/3,600 R.P.M. available from normal 50/60Hz induction machines. At these very high speeds small induction motors can produce extraordinary amounts of power - a 1H.P. motor may only weigh a pound or two.

Although commutator motors can operate in this speed range, commutator and brush wear is a problem and they cannot give the long trouble-free service needed for continuous use in industrial applications.

6.1.2 Commutator motors

In contrast to induction motors, commutator motors are relatively easy to control over a wide range of speeds.

The factors which affect speed variation have already been covered in section 4.7 of this book and this section is mainly concerned with methods of arranging the control equipment.

For most applications, variation of speed is not the only requirement. It is also necessary for the speed, once set, to be reasonably independent of the load placed on the motor. This is best met by a control of the armature voltage of a commutator motor fitted with either a permanent magnet field, or a shunt field separately supplied at its rated voltage.

Small speed variations are possible by the use of a variable resistor in series with the armature. However, as soon as a load is placed on the motor, the increased armature current adds to the voltage drop in the resistor and decreases the voltage available at the armature. This results in very poor speed regulation and this method is only really suitable for speed ratios of perhaps 1.5:1 and with the motor driving a reasonably constant load. For wider speed ratios and variable loads the variable voltage source must have a low impedance so that the voltage applied to the motor does not vary appreciably as the current taken by the motor changes.

For small low voltage motors the simplest method is to use a power transistor to stabilise the motor voltage. A suitable arrangement is shown in figure 6.2. In this circuit Q_2 ensures that Q_1 is always turned on just enough to keep Q_2 emitter (i.e. the output voltage) just less than the voltage at the slider of the potentiometer RV1. This circuit has a very low output impedance and the largest source of unwanted variation is the change in input voltage as the motor current varies.

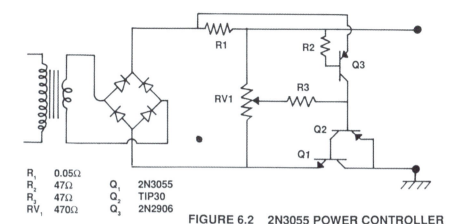

R_1	0.05Ω		
R_2	47Ω	Q_1	2N3055
R_3	47Ω	Q_2	TIP30
RV_1	470Ω	Q_3	2N2906

FIGURE 6.2   2N3055 POWER CONTROLLER

The circuit as shown is suitable for continuous currents of up to 4A and the maximum possible output current is limited to about 12A by R_1 and Q_3. Q_3 is normally off but it starts to conduct when the voltage drop across R_1 reaches 0.6V. This shuts down the output stage and prevents the current rising any further.

The transistor types are not critical and, provided current ratings are not exceeded, alternative types can be substituted without problem. The difference between the supply voltage and the motor voltage appears across Q1. Since this is passing the full motor current, the dissipation is quite high and it must be mounted on an adequate heat sink to get rid of the heat. The size of the heat sink depends on the regulator rating. If in a particular case the motor current is, say, 2A and the maximum voltage dropped across the regulator is 12v then some 24 watts will be dissipated. The 2N3055 can dissipate this amount of power, with an ample margin in hand, when it is bolted to metalwork as hot as 100 degrees C. This may be any piece of metal which is large

enough and thick enough or a commercial finned heat sink. If a commercial heat sink is used one with a rating of about three degrees C per watt is needed. If local metalwork is used the minimum size depends on the material and the thickness. As a general guide 100 square inches/600 square cm of 0.04in/1mm aluminium sheet is sufficient. If in doubt sprinkle a few drops of water on the transistor metalwork (the case, not the terminations!). So long as the water doesn't boil it's cool enough. Don't forget that this heat sink is connected to the negative output terminal. It is best to ground this terminal but if this is not possible then the heat sink should be insulated and made safe from accidental short circuits. It is possible to buy mica washer mounting kits to insulate the power transistor from its heat sink but these are an additional barrier in the cooling path and it is better to avoid them if possible.

For currents up to 1.5A the National LM317T or Texas Instruments LM317KC integrated circuit regulator is a very

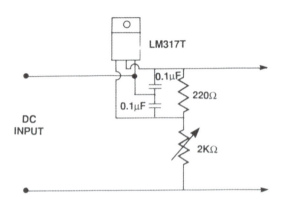

FIGURE 6.3   LM317T POWER CONTROLLER

useful device. This is a variable voltage regulator that can be set to any output voltage from 1.2v to 37v. It has the great advantage that it has built-in current limiting and thermal protection. This means that it will withstand accidental short circuits and if, for any reason, it overheats it will automatically reduce its output current to a safe value.

The method of use is shown in figure 6-3. If the LM317T is to be used near its maximum current rating a fairly large heat sink will be needed – again about 100 square inches/600 square cm. However, some risks can be taken with this if the LM317T is only operated at high dissipation (i.e. high current at *low* output voltage) occasionally, as it will automatically protect itself by reducing its output curren if it gets too hot.

Both of these methods are convenient for small, low voltage motors. With larger motors, however, the power to be dissipated in the heat sink becomes excessive. In addition to this, high voltage output may be needed. High voltage transistors are readily available but they are much less forgiving devices than their lower

voltage counterparts and will only function reliably if their operating conditions are closely optimised about a particular application.

Fortunately these higher voltage, higher power applications are well suited to thyristor (often called an SCR – short for Silicon Controlled Rectifier) or triac controllers. These are a sort of silicon switch – a very low level control signal can turn them "on" but then they will stay on as long as power flows through them. This makes it very difficult to use them from D.C. supplies because there is no easy method of turning them off. However, they are ideally suited to A.C. supplies. With an A.C. supply the voltage passes through zero twice on each cycle of the mains frequency and this automaticlaly resets the silicon switch ready to be triggered on again by the control signal.

A thyristor can only pass current in one direction so that it is necessarily off for at least half of the time. A triac can pass current in both directions and so can be "on" for most of the time. Both devices have their advantages but, in general, the thyristor is less fussy about its operating conditions than the equivalent triac and so is often preferred for general purpose applications.

Triacs and thyristors are the main control element in domestic light dimmers and also for speed control in portable electric tools. Frequently a sub-assembly can be liberated from one of these and used to control other motors.

A typical arrangement is shown in figure 6-4. This is a thyristor controller driving a universal series wound motor. The circuit relies on the fact that, when power is removed from a rotating series wound motor, it generates a small voltage directly proportional to speed. Although no field current is flowing, there is sufficient residual magnetism left in the stator iron to enable the motor to behave as a permanent magnet D.C. generator with an extremely weak field. The value of this generated voltage is balanced against the voltage from the speed setting potentiometer VR1. The difference between the two voltages is used to control the time delay before the thyristor is triggered on again. Because the thyristor always switches off at the same time in the supply frequency cycle the delay before it triggers "on" controls the length of the "on" period which in turn controls the average voltage applied to the motor.

With the speed control at maximum the thyristor is fully conducting on alternate half cycles and the motor will run at about two thirds of its full voltage speed. S_1 is provided to short out the thyristor when full voltage and speed is needed.

6.2 Electric Braking

6.2.1 Induction motor braking

Induction motors can be braked effectively by passing a D.C. current through one or more of the windings ("D.C. Injection" braking). As the conductors in the squirrel cage rotor cut the stationary stator field, large currents are induced which initially exert a strong braking force on the rotor. However, at low speeds the strength of the induced current falls and with it the braking force, eventually reaching zero braking force at zero speed. The actual curve of braking

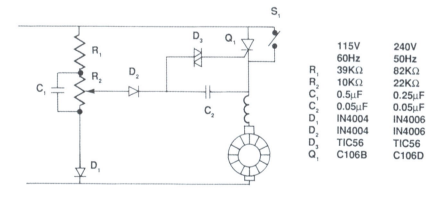

	115V	240V
	60Hz	50Hz
R_1	39KΩ	82KΩ
R_2	10KΩ	22KΩ
C_1	0.5μF	0.25μF
C_2	0.05μF	0.05μF
D_1	IN4004	IN4006
D_2	IN4004	IN4006
D_3	TIC56	TIC56
Q_1	C106B	C106D

FIGURE 6.4   5 AMP THYRISTOR CONTROLLER

111

force versus speed is very similar to a mirror image of the normal torque/speed curve of the motor so that the previous zero torque output point at synchronous speed now occurs at the origin (Figure 6-5) i.e. the motor is trying to behave as an induction motor with a synchronous speed of zero r.p.m!

The amount of braking torque depends on the current through the windings – an amount roughly equal to full load current is often sufficient and has the advantage of little increase in winding temperature. The power dissipated in the windings is no higher than in normal operation. A considerable amount of braking energy is dissipated in the rotor but this is well able to withstand high temperatures and loses most of its heat to the airstream through the motor.

More braking can be obtained by increasing the D.C. to two or three times full load current but beyond this magnetic saturation of the iron circuit limits the increase. The peak braking force that can be achieved is well above the normal starting torque and, of course, the torque reaction is in the opposite direction. Because of this make very sure the motor is properly anchored before making any braking experiments – a heavy motor cavorting round the workshop can do a lot of damage!

At these higher currents each stop puts additional heat into the windings so that frequent braking can lead to overheating.

The braking power can be obtained from a transformer and rectifier or, less efficiently but more conveniently, from a dropping resistor and rectifier. Because the D.C. resistance of the windings is much less than their A.C. impedance the voltage required is quite low – only about one tenth of normal. Suitable arrangements are shown in figure 6-6. The dropping resistor can be a small electric fire element for large motors or an appropriately sized light bulb for

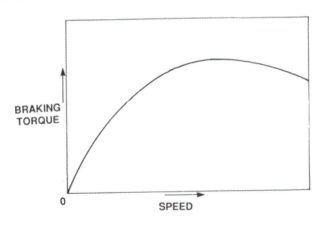

FIGURE 6.5   D.C. INJECTION BRAKING
TORQUE/SPEED CURVE

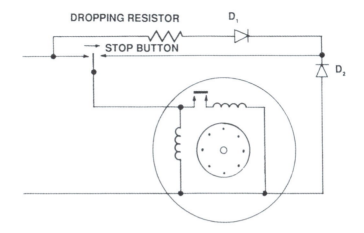

FIGURE 6.6   D.C. INJECTION BRAKING

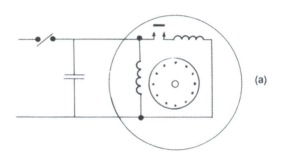

(a)

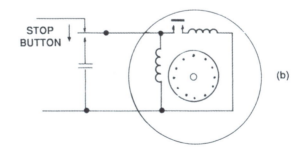

(b)

FIGURE 6.7   REGENERATIVE BRAKING

113

smaller machines. The rectifier system in figure 6-6 uses a half bridge instead of the more common full bridge as it wastes less power in the dropping resistor. On the positive half cycle current flows through D1 into the winding and stores energy in the magnetic field. On the negative half cycle no current flows through the dropping resistor but, as the magnetic field starts to collapse, it develops a negative voltage overshoot which causes almost the original amount of current to continue to flow, this time through the "free wheeling diode" D2. In a loss-free system the average current through the winding would be twice the average current through the dropping resistor — in practice about one and a half times is possible. This is not something for nothing but a full explanation is beyond the scope of this book.

To disconnect braking current when the motor is not in use it is usual to use a contactor with separate "press to start" and "press to stop" buttons. The stop button is arranged so that braking current is only applied while it is held down. As soon as it is released all power is removed from the motor. This is very convenient if full braking is only required occasionally. A quick touch on the button will allow the motor to stop normally. If the button is held down, full braking is brought into play.

This braking system is not at all critical in operation and will work on practically any type of induction motor. There is normally little to be gained in passing the braking current through more than one winding — the main winding in a single phase machine or any convenient pair of terminals on a three phase machine.

An alternative method of braking is regenerative capacitor braking. If a squirrel cage motor is mechanically driven and connected to a capacitative load it can operate as an induction generator. With the right value of capacitor chosen in relation to the motor characteristics large circulating currents can flow and produce braking characteristics initially similar to D.C. injection braking. A typical arrangement is shown in figure 6-7a. Although this is a simpler arrangement than D.C. injection it is more dependent on the characteristics of a particular motor and rather large values of capacitor are needed — 50μF or so for a 240V ½ H.P. motor. The initial braking torque is large but it falls off very rapidly as the speed drops so that the improvement in overall time to coast to a standstill is limited. The alternative arrangement in figure 6-7b disconnects the capacitor in normal operation and only brings it into circuit for the braking mode. Although kinder to the capacitor it is less effective as a braking system because there is now about a one second delay while the braking torque builds up to its maximum value.

Motors can also be obtained with an electromagnetic disc or drum friction brake mounted on a shaft extension on the non-drive end of the motor. These are convenient and effective devices and have the advantage that braking torque is maintained down to zero speed. However, they are comparatively rare and difficult to obtain for one-off applications. A reasonable alternative is the use of an electromagnetic clutch either fitted to the motor shaft or at some convenient point in the drive train. Devices of the type shown in figure 6-8 are available in a wide range of torque capability and coil voltage. These are primarily intended for use as clutches to connect or disconnect two rotating shafts but are equally suitable for use as brakes.

6.2.2 Commutator motor braking

Commutator motor regenerative braking is even simpler than induction motor regenerative braking. Provided the motor field is maintained, it is only necessary to short circuit the brushes, either directly or through a current limiting resistor. With the field maintained the commutator motor now acts as a D.C. generator. Large currents flow, and the braking energy is dissipated as heat in the armature winding resistance and the external resistor if fitted. The braking torque is directly proportional to current and, since this falls with speed, the braking torque reduces at low speed, falling to zero when the armature is stationary.

Maximum braking torque is achieved with a direct short circuit and this is normally permissible with small motors up to about ⅛ H.P./100W. With larger motors excessive peak currents can occur if the brushes are directly short circuited at high speed and this can result in high brush and commutator wear. In addition to this, in permanent magnet machines, very high peak armature currents can partially demagnetise the field magnets. A peak current of up to about four times full load current is usually safe. The minimum value of resistor required depends on both the motor characteristic and the initial speed but a value of three to five times the armature resistance is fairly typical.

With permanent magnet machines the right field conditions for braking are automatically available. With shunt wound machines, field current must be maintained in its normal direction during braking and it is usually quite simple to arrange this. Series wound machines have a special difficulty. The normal series field is provided by current flowing through the field into the armature. This field must be maintained *in the same direction* when current flows *out* of the armature during braking. This means that, in the braking mode, the connections to the series field must be reversed. The reversal is essential because if the original direction were maintained just sufficient initial braking current would flow to cancel the residual magnetism in the field iron circuit. The armature current would then drop to zero and from then on no further braking would be available. The braking characteristic is related to the residual magnetism direction at the time power was removed from the motor – the braking current must reinforce this.

The braking characteristic is independent of the polarity of the voltage originally applied to the motor so this system works equally well on universal motors working from A.C. supplies.

6.3 Workshop applications of electric braking

The main use of braking in the workshop is to reduce the stopping time of machines used in stop start operation. High speed machinery for wood, particularly saw benches and planers, take an inordinately long time to come to a halt after power is removed. Electric braking can substantially reduce this but don't forget to let friction finally halt the machine before allowing your fingers near the blade.

Drill presses and milling machines

Fig. 6.8 *Electromagnetic friction clutches*

normally stop sufficiently rapidly but lathes operated at high speeds with heavy chucks can take a long time to coast to a standstill. Electric braking works fine here in light duty service with only a few stops in quick succession. However, it is unsuitable for the large number of stops and starts per hour needed for some types of production operation, as frequent starts and stops can lead to overheating. Production lathes intended for this sort of service are normally provided with a hand operated clutch between the motor and the main spindle. With this system the motor runs all the time and the workpiece is stopped and started with the clutch.

CHAPTER 7

Generators

7.1 D.C. Generators

Almost any commutator motor will act as a D.C generator (often called a Dynamo) if driven at a suitable speed and arrangements are made to initiate and maintain the stator field.

A permanent magnet machine always has full field present and will generate a no-load output voltage directly proportional to speed. The output polarity depends on the direction of rotation and will reverse if the direction of rotation reverses. If driven at constant speed, the generator behaves as fixed voltage source in series with the armature resistance so that the voltage drop across the armature resistance causes the output voltage to fall when the generator feeds an electrical load.

The only method of maintaining the output voltage constant as the load increases is to increase the speed and this is usually very inconvenient. Because of this permanent magnet generators are rarely used for power generation.

The shunt wound machine has more favourable characteristics because the output voltage can be controlled by varying the field current. This makes it possible to deliver a constant output voltage over a wide range of shaft speeds and electrical load.

It is normally necessary for a generator to supply its field current from its own generated output. This can be a problem because, unless at least a small amount of field is present at start up, it will fail to generate. A very small amount is sufficient because, as soon as the generator produces some output, it increases the field and the output rapidly builds up to normal level.The residual magnetism that remains in the iron of the field circuit after power has been removed is normally sufficient. This is only true if power has been previously applied to the field. If a machine has been taken to pieces and the armature removed from the the field tunnel the higher magnetic reluctance caused by the large air gap will destroy most of the residual magnetism. When reassembled, it will fail to self generate until the residual magnetism has been restored by momentarily applying normal voltage to the field.

It will also fail to self generate if the direction of rotation is reversed because the output polarity is now reversed and the resulting field current cancels instead of aiding the residual field strength. To restore self generation the field connections must be reversed and the residual magnetism restored.

A typical shunt wound generator is the 6 or 12 V dynamo that used to be common in automobiles. These start to produce useful output at less than 1,000 R.P.M. and are fitted with a regulation system which controls the output voltage to a preset level over the speed range from about 1,000 to over 5,000 R.P.M. The regulator is essential because, even at constant speed, the output voltage of a shunt wound generator is inherently unstable. Once the speed is high enough for the machine to self generate, any increase in output voltage causes a corresponding increase in field current which again increases the output voltage...... This process will continue with the voltage increasing until the iron circuit of the field is saturated and the field strength cannot increase further. Once the field is saturated it behaves as a permanent magnet machine and the output voltage then rises directly with speed. In practice

the power dissipated in the field coils becomes excessive before full magnetic saturation is reached and the regulator is used to maintain constant output voltage over the useful speed range by field currents varying from about 15% to 90% of saturation value.

This runaway voltage characteristic only applies to a shunt wound machine driving either no load or a light load. Provided the speed is maintained within reasonable limits a starter battery connected directly across the output will exert a strong stabilising influence. If the output voltage tries to rise above the nominal battery voltage heavy charging currents flow. However, the field voltage is prevented from rising significantly and the excess voltage is dissipated in the internal resistance of the armature. Over a small speed range the voltage is sufficiently constant to drive simple lighting circuits. For more demanding applications a voltage regulator is essential.

A simple constant voltage regulator circuit is shown in figure 7- 1. The power

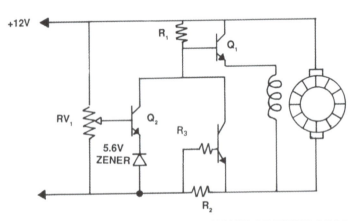

FIGURE 7.1   VOLTAGE REGULATOR

118

transistor Q1 is turned on by R1 and passes current into the field winding. Q2 is off until the voltage at the slider of RV1 exceeds the zener voltage of zener diode D1. When this happens Q2 starts to conduct, diverting some of the current which R1 supplies to Q1. This then reduces the current that Q1 can pass through the field coils, stabilising the output voltage at a level just high enough to allow Q2 to conduct.

Q1 and Q2 stabilise the output voltage at a fixed level determined by the setting of VR1, but if connected to a completely discharged battery excessive current would flow. Q3 is added to limit the maximum current. Q3 is a silicon transistor which does not pass current until the base voltage is about 0.6v higher than the emitter. At normal currents Q3 is off and does not affect the output voltage. However, if the current rises high enough it will eventually develop 0.6v across the current-sensing resistor R2 and from that point onward Q3 will limit any further increase in current by progressively reducing the output voltage.

In the same way that commutator motors can be operated at different speeds by choice of armature voltage, D.C. generators can be operated over a range of output voltage provided the field current is set at a suitable level. A 12V volt dynamo rated for full output at 1,000 R.P.M. will quite happily produce 24V at 2,000 R.P.M. if a dropping resistor or other suitable method is used to keep the field current within its normal rating. Providing the armature is mechanically suitable for operation at the higher speed, the machine is not overstressed and will deliver its full rated current at twice its rated low speed voltage.

If reduced output voltage is needed, and it is also desirable to operate the dynamo at low speed, then a 2:1 improvement can be obtained by reconnecting the two field coils in parallel instead of their normal series arrangement. A twelve volt dynamo reconnected in this way will develop six volts at about half the rated speed of the original twelve volt connection. There are penalties attached to this low speed configuration. The first snag is that the current rating is not increased so that halving the output voltage also halves the maximum output power. Also, because the power lost in the resistance of the armature conductors is unchanged it now forms a much larger fraction of the output power and the efficiency will be lower.

Although a series wound machine can operate as a generator, the field strength is determined by the output current and cannot easily be separately controlled. It is only really suitable for applications where the machine is suddenly called upon to produce the maximum power that it is capable of delivering. Regenerative braking is one such application and has already been described in section 6.2.2.

Series windings are sometimes added to shunt wound machines to modify the output characteristics. A small series aiding winding will increase the field strength as the armature current increases and the amount can be chosen to cancel the voltage drop in the armature resistance. The correction will not be perfect but it can make the voltage regulation task easier by greatly reducing the change in field current needed to compensate for load changes.

D.C. generators for arc welding need what is usually called a "drooping voltage characteristic". This is an output voltage characteristic which falls (droops) rapidly as the arc current increases. This is to maintain a stable arc current in spite of variations in arc length. This can be achieved by adding series opposing turns to the main shunt field. Now, any increase in arc current decreases the generator field strength and the number of turns in the series field can be chosen to give the optimum amount of "droop" in the output voltage.

7.2 A.C. Generators

A.C. generators (usually called alternators) cover a number of basic types optimised for different applications. Perhaps the most familiar type is the alternator that has now replaced the dynamo in most transport vehicles. An alternator of this type is shown in figure 7-2. In this case the rotor carries the field winding and power is taken from the three phase winding on the stator. Most machines described so far have carried their field winding on the stator and their main winding on the rotor and alternators can be built in this way. However, this reversal of functions is commonplace in alternators as it has the advantage of both simplifying and reducing the cost of the slip ring and brush assembly which feeds current to the rotor. This is because the power fed to the field winding is typically less than 10% of the main winding power, and in addition, two slip rings are sufficient - a rotating main winding may need three or more slip rings.

The alternating voltage in the main stator winding is generated as the stationary stator conductors cut the rotating magnetic field generated by the current fed to the field winding. In principle the rotor can have any number of pairs of poles. For fixed frequency

Fig. 7.2 *A vehicle alternator*

applications the number of poles is usually chosen to permit operation at a convenient shaft speed. Two pole fields are often chosen because this permits the highest shaft speed for a given output frequency and this is usually the minimum size and cost arrangement. At the other end of the scale very large numbers of poles are occasionally chosen to permit direct drive from a low speed prime mover.

In the case of the vehicle alternator all the output power is rectified and used to charge the battery and operate the vehicle equipment. In this application frequency is unimportant and the number of rotor poles can be chosen on the basis of manufacturing convenience and cost. The rotor shown in figure 7-3 has six poles formed by the two sets of three teeth on two interleaved cup-shaped pressings. All six poles are energised by a single bobbin-wound coil

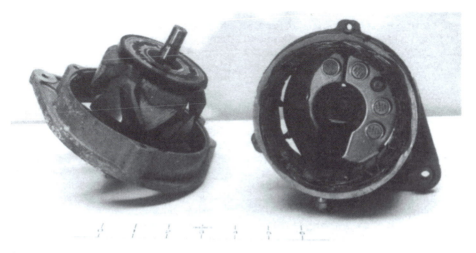

Fig. 7.3 *The rotor of a vehicle alternator*

surrounding the rotor shaft inside the two cups. This results in a very simple and rugged rotor assembly that can operate at the high speeds and temperatures necessary in vehicle service.

The number of poles chosen for the rotating field is not related to the number of phases in the main stator winding. In this case there happens to be a simple 6:3 ratio but this is in the nature of a coincidence. Depending on other constraints perhaps two to twelve poles and one to three phases would have been equally valid choices.

A three phase stator is chosen in this case as it permits more efficient use of the stator copper and iron than a smaller number of phases. The three phase main winding feeds six diodes connected as a three phase full wave rectifier. The output from this feeds the vehicle load and, via the voltage regulator, the alternator field winding.

In this alternator/rectifier combination the rectifier carries out the function performed by the commutator in a dynamo and the overall behaviour is almost identical to the shunt wound D.C. machine discussed in paragraph 7.1. It can in fact use the same type of voltage and current regulator.

The A.C. output of this type of alternator can of course be used to power a load directly. In small installations, however, it is difficult to make effective use of three phase power and, although slightly less efficient, a single phase output is more useful.

For small machines a permanent magnet rotor can be used in place of the rotating wound field. It is no longer possible to regulate the output by controlling the field current but a fairly stable output can be achieved into a fixed load by special design of the single phase alternator winding. This is designed to have high leakage inductance and a higher than normal iron loss. The iron loss and the reduction in output due to the leakage inductance both increase with frequency. Over limited range these increasing losses can be arranged to offset the normal increase in alternator output with speed and deliver a reasonably constant output.

The above types of alternators generate output as a result of relative rotation between the field magnet and the output windings. An alternative type known as an inductor alternator allows both elements to be stationary. It generates its output by rotating a laminated soft iron rotor which steers the field flux through alternate paths in the output winding.

The flux paths of an inductor alternator are shown in figure 7-4. The two field windings induce a North pole in the upper half of the stator and a South pole in the lower half. The direction of the flux through the two A.C. output windings is controlled by position of the laminated soft iron rotor.

In the left hand diagram this flux direction is from right to left. The right hand diagram shows that a 90 degree rotation of the rotor redirects the flux to give a complete flux reversal through the two output windings. Each further 90 degrees of mechanical rotation results in successive 180 degree changes in flux direction, going through two complete cycles of 360 electrical degrees in one full rotation of the rotor. This means that this is the equivalent of a four pole single phase alternator.

To simplify the illustration the field and main windings are shown as four separate windings each encircling part of the laminated stator. In fact the only useful part of these windings is the parts of the turns that lie within the stator. All the induced voltage appears in this part of the turn. The remainder of the winding which lies outside the stator tunnel simply serves to connect all the

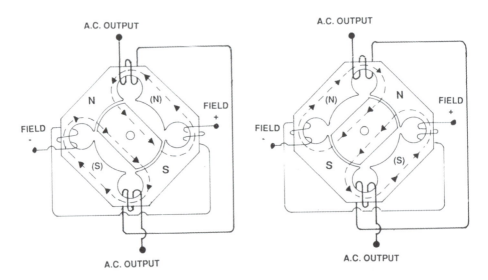

FIGURE 7.4   INDUCTOR ALTERNATOR

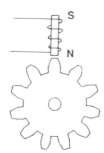

FIGURE 7.5   INDUCTIVE PICK-OFF

useful parts of the windings in series. Because of this, this outside return path of the winding can take any convenient route. In many cases this results in a saddle shaped coil with the active part passing through pair of diametrically opposite stator slots, rather similar to the winding that surrounds the field pole piece of D.C. machine but this time embracing two pole pieces instead of one.

This type of alternator is very useful for small machines operating at very high rotor speeds. No slip rings are necessary and the simple laminated rotor can withstand very high centrifugal forces.

The alternator described above is a bipolar type because, during one cycle of the output frequency, the magnetic flux density in the output winding changes from a maximum value in one direction to a maximum value in the other direction. This makes full use of the maximum flux swing possible and permits the design of alternators which approach the efficiencies possible with rotating field machines.

An even simpler type is the homopolar induction generator. This no longer attempts to use the full bipolar flux swing but contents itself with a flux variation from a maximum value in one direction to a lesser value in the same direction. This principle can be used to generate useful amounts of power at fair efficiency (in World War II the early British airborne radars were powered by 1600Hz aircraft alternators of this type). However, by far the most common use is in the form of inductive pick-offs used to generate low level electrical signals to monitor shaft speeds and relative positions.

In this application, simplicity and small size are the primary aim. These pick-offs usually drive electronic circuits so extremely low powers (microwatts) are enough and efficiency is unimportant. A common use is to generate a small A.C. voltage whose frequency is a simple multiple of shaft speed. An electronic circuit then indicates shaft speed by measuring the frequency of this voltage. This low level voltage can be generated very simply by a small cylindrical magnet surrounded by a coil and positioned with one end of the magnet near a toothed iron or steel wheel (figure 7-5). A special tooth form is not necessary and it is often possible to use the teeth of one of the gears already present. As the tooth approaches the tip of the magnet it reduces the total reluctance between the North and South poles. This results in a small increase in flux density and this change is enough to generate the small fraction of a volt needed by the electronics.

CHAPTER 8

Installation and Protection

8.1. General

This chapter is mainly concerned with the installation and protection of fractional horse power motors in small workshops. Other parts of this book cover a much wider range of motor types and applications but the range is too varied for all aspects to be covered in a general chapter. However, many of the problems are common and similar techniques can be used where appropriate.

8.2 Installation

If a standard lathe or drill press is used which already includes arrangements for the motor mounting and drive shaft coupling then the mechnical require-ments are straightforward. The whole assembly needs to be mounted on a sturdy bench and protected against one of the small workshop's more common hazards – condensation. If the strength of the bench and the mountings are chosen in relation to the size and weight of the machine this will normally be more than sufficient as far as the motor is concerned. Condensation is a different matter. Machine surfaces can be protected against rusting by liberal

applications of grease or oil but motor survival is totally reliant on the varnish coating applied by the manufacturer to protect the winding against ingress of moisture. A great deal of effort has been expended on the development of the special resins used for this purpose and modern resins give a remarkable degree of protection. Nevertheless if long and troublefree service is to be obtained, it is prudent to take reasonable precautions to minimise condensation and to prevent oils and cutting fluids reaching the windings.

Many workshops have poor thermal insulation and are only heated for a few hours each day. This is a recipe for frequent, heavy condensation. The worst condensation occurs when a cold motor is in contact with warmer, moist air. The cold motor attracts a downcurrent of the warmer air which rapidly deposits a film of water where it is least welcome. If this is a problem, a sheet of polythene or even an old newspaper loosely draped over the danger area when the machine is out of use will give a surprising degree of protection. Once the downcurrent is interrupted, moisture can only

Fig. 8.1 *A simply-made cradle*

condense out of the local stagnant air and this rarely contains enough moisture to be a problem.

8.3 Motor mounts

Most fractional horsepower induction motors come provided with foot mountings (figure 2-2) and this is ideal for the commonly used belt drive to the load. It is less suitable for gear drives or splined couplings but, since most motors are held together by four long bolts securing the end bells to the stator laminations, it is not too difficult to improvise a flange mount (figure 2-4) secured by adaptors to these four bolts.

Adapting a purely cylindrical or flange mounted motor to foot mounting may not be as simple. Some modern motors have tapped holes already provided in the stator casing and the motor manufacturer will supply castings or heavy duty pressings that can be

bolted on to form a foot mounting. However, in many cases a cradle is the best solution. Figure 8-1 shows an easily made cradle. It consists of two pieces of standard Dexion angle each with a semicircular cut-out to fit the motor. In this particular cradle the two pieces are held together by a pair of mild steel plates welded to the ends of the angles. However, welding is not essential – the ends of the angle can simply be bent round and riveted together or the two vertical sides can be secured by two long bolts passing through two large diameter spacers which maintain the sides at the right separation. In either case the motor is retained in the cradle by a large worm drive Jubilee pipe clip. This is cut in two and bolted or riveted to the end plates. The horizontal parts of the mounting can face inwards or outwards and can have the normal Dexion holes elongated into long slots to make provision for adjustment of belt tension. The particular version shown in the illustration has one angle turned inwards to permit the cradle to be close to the drive shaft end of the motor.

With most belt drives some means of adjusting belt tension is necessary. With fixed drive centres this can be achieved by an adjustable jockey pulley bearing against the inside or the outside of the belt path about half way between the drive centres. However, in most applications it is simpler to adjust the drive centre distance for proper belt tension by moving the motor. If single speed pulleys are fitted adjustment will rarely be needed and slotted hold downs as in the cradle described in the last paragraph are probably sufficient. These slots provide the free movement necessary to adjust belt tension but they have the disadvantage that they do not ensure that the motor spindle stays accurately at right angles to the plane of the belt loop. It is only too easy to finish the belt adjustment with the correct belt tension but with enough skew to result in rapid belt wear.

For maximum belt life both the drive shaft and the driven shaft must be accurately at right angles to the plane of the belt loop and the pulleys themselves must lie in that same plane. If the pulleys are of equal width this is easily checked by laying a straightedge or a taut string along the sides of the two pulleys and as near the shafts as possible. If the drive is correctly aligned the rim of each pulley will touch the straightedge at two points. If the pulleys have different overall widths then the straightedge should clear the narrower pulley by half the difference in width.

A much better method of belt tension adjustment is to swing the whole motor about an offset hinge pin. This allows the tension to be adjusted without upsetting the shaft and pulley alignment. A simple way of arranging this is to bolt a foot or cradle mounted motor to one end of a rectangular baseplate. The further end of the base plate is secured to the machine by a pair of ordinary household door hinges. For most purposes wood is a perfectly satisfactory material for the baseplate and, if painted a suitable colour, doesn't look too out of place. If the imaginary line joining the motor shaft centre to the hinge pin centre is at right angles to the second imaginary line joining the motor shaft centre to the driven shaft centre than a large change in centre distance can be obtained from a small angular movement at the hinge (Figure 8-2a). For vertical shafts this is usually the best arrangement.

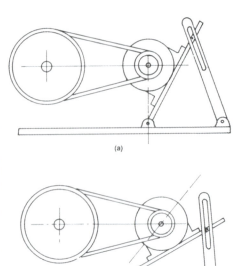

(a)

(b)

FIGURE 8.2   HINGE TYPE MOTOR MOUNT

For horizontal shafts there is some advantage in the set up of figure 8-2b. In this the shaft/hinge pin line is tilted back about 45 degrees. This allows the weight of the motor to provide part or all of the required belt tension. If the weight of the motor is sufficient this results in a self adjusting drive. However, in most cases more tension will be needed and the slotted tension adjusting arm shown in figure 8-2a will be necessary .

8.4 Drive arrangements

The great majority of workshop applications use belt drives to couple the motor to the load. Belts have the great virtue that they are very tolerant of slight misalignment between motor and load. The belt length can be chosen to locate the motor in the most advantageous position and, apart from timing belts, can be arranged to slip on heavy overload to protect motor and machine from damage.

Belt drives are mainly suitable for modest reduction or step-up ratios. In most workshop applications 3:1 is about the useful maximum with reasonably sized pulleys and centre distances. For some applications about 15:1 is possible with POLY-V belt drives and over 20:1 with modern synthetic flat belts. However, these are specialised applications with very large low speed pulleys.

For maximum power transmission capability the belt speed should be kept high – usually in the speed range 1,000 to 5,000 feet per minute/5 to 25 metres per second. This may not be possible in some applications because of limitations in the size of the pulleys. This does not prevent the use of belts but it means using stronger belts of larger cross-section.

Maximum power transmission capability occurs at 1:1 ratio with 180 degree wrap round each of the two equal sized pulleys. At this ratio the distance between shaft centres is unimportant as it does not affect the angle of wrap. However, with larger ratios, the angle of wrap round the smaller pulley decreases rapidly as the centre to centre distance is shortened (see figure 8-4). If possible, the shaft spacing should be large enough to permit 120 degree wrap round the small pulley as this will achieve most of the power transmission capability of the drive. Wrap angles down to about 90 degrees can still handle about two thirds of the drive capability but at angles less than this the performance falls off rapidly.

The main types of belt drive are:-

Vee belts
flat belts
round belts
timing belts

Some of these belts are shown in figure 8-3.

Vee belts are the most commonly used types and are available as endless belts in a wide variety of lengths and cross sections. All types have a wedge angle of 40 degrees and it is the wedging action of the sides of the belt against the sides of the pulley that enables the vee belt to transmit more horsepower per inch of width than the older types of flat belt. Most of the belt is canvas covered rubber which provides the wedging action. The tensile strength of the belt is provided by a number of cords buried in

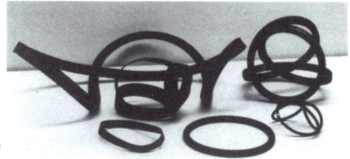

Fig. 8.3 *Various forms of drive belt*

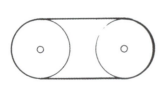
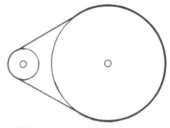

FIGURE 8.4   PULLEY WRAP ANGLE

the rubber just below the flat top of the V (figure 8-5). When bent round a pulley the inside part of the V is squashed and travels at a lower speed than the flat outer portion which is slightly stretched. The cords runs through the neutral axis of the belt i.e. the part of the belt which is neither stretched or squashed as the belt is bent. When stretched round a pulley it is the position of this neutral axis or "pitch-line" on the pulley that determines the effective pulley diameter. This is the diameter to be used when calculating drive ratios. It lies about one third of the belt thickness below the flat top.

In large diameter pulleys the sides of the belt groove are set at 38 degrees – just less than the nominal 40 degree belt wedge angle. This ensures that the main wedging action is biased towards the outer edge of the belt close to the pitch line.

When wrapped round a small diameter pulley the reduction in length of the inner part of the belt section makes it bulge out sideways and reduce the belt wedge angle. To compensate for this smaller pulleys are manufactured with reduced groove angles – about 32 degrees for the smallest practical pulleys and 35 for intermediate sizes.

Ample clearance must be provided between the inner surface of the belt and the base of the pulley groove. As the belt and pulley wear in service the belt drops deeper into the pulley groove but must never touch bottom. Should this happen the wedging action ceases and, unless the drive is very lightly loaded, severe belt slip results. In a properly adjusted drive the upper part of the pulley groove walls should wear clean and lightly polished but the groove base should be dull, matt and usually dirty. A clean polished groove base is clear sign of a slipping belt that has bottomed. When this happens both belt and pulley wear rapidly and it may be necessary to replace both items.

FIGURE 8.5   VEE BELT
CROSS SECTION

FIGURE 8.6   POLY-V BELT
CROSS SECTION

This problem is usually the result of a badly tensioned belt. To avoid rapid wear belts must always have sufficient tension to avoid slip in normal service. A properly tensioned belt will feel springy and vibrate if struck sharply with the side of the hand. If firm pressure is applied to the middle of the unsupported part, the belt should not deflect more than its own depth before strong resistance to further movement is felt.

For powers that are higher than can conveniently handled by a single V belt, the power-handling capacity can be increased by the use of matched sets of two or more belts in multi-grooved pulleys.

A variant of this which is particularly useful if high ratios are needed is the POLY-V belt section shown in figure 8-6. In this from two to twenty rubber vees are moulded beneath a layer of load-carrying steel or textile cords. The multiple vees provide a grip comparable to a V belt of the same width but of much greater section depth. The section depth of the POLY-V belt is not much greater than that of a flat belt and this makes it possible to use it with small diameter pulleys for high ratio drives capable of operating at high shaft speeds. Many domestic washing machines use this type of belt to provide a reduction radio of about 1:12 from a high speed commutator motor to the main wash/spin drum.

Flat leather belts were originally the mainstay of industrial power transmission systems but their place has mainly been taken by V belt systems which, because of their wedging action and high strength cords, can handle a much greater power per inch of pulley width. A few leather old faithfuls survive in ancient machines in small workshops (mine included) but they are no longer in serious use and replacements are practically unobtainable. The wide thin-walled pulleys used for leather belt transmission are rarely suitable for conversion to standard V belts but are an excellent basis for conversion to a POLY-V drive which will then outperform the original drive by a considerable margin. The main current use of flat belts is in the form of endless belts made up of a very thin flat layer of synthetic fibre cords faced either side with a rubber or polyurethane driving surface. Although very strong, the belt thickness is typically only 0.040"/1mm – much lighter and more flexible than similar width V form belts. This makes them suitable for very high speed operation – up to 10,000 feet per minute/55 metres per second. They can also be used successfully with very small pulleys which permits large single stage ratios – over 20:1.

There is no wedging action with this type of belt so, at constant speed, the horsepower that can be transmitted per inch of belt width is less than V belt systems. However, most of this can be recovered by operating at the higher belt speeds that are possible with this form of construction.

An interesting facet of flat belt operation is its apparently perverse behaviour when running on plain pulleys not fitted with flanges. It would be reasonable to expect that if the running surface of pulley were made slightly concave the belt tension would cause it to slide to the bottom of the depression and centre the belt on the pulley face. This does in fact happen if the belt slides over the surface of a stationary pulley or if it is operating with insufficient tension and slipping badly. However, if the belt is properly

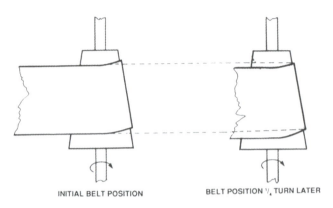

tensioned and operating normally it will promptly climb over the edge of the pulley and firmly resist any attempt to guide it back to the centre of the pulley face.

If, however, the pulley face is "crowned", i.e. made slightly convex, the behaviour is reversed – a strong self-centering action in normal drive but sliding off one side if the drive is overloaded and the belt slips. The reason for this can be seen in figure 8-7. In this a wide flat belt is shown running on a strongly coned pulley. The belt tension forces the belt to lie along the surface of the cone at the furthest extension of the belt i.e. at the right hand side of the figure. This distorts the entry of the belt and forces the part of the belt just arriving at the pulley to make contact slightly further up the face of the cone. The effect is cumulative and with the rather extreme cone angle shown in figure 8-7 a few revolutions will be enough to cause the belt to climb right up the cone and over the pulley edge.

The effect can easily be demonstrated with a wide rubber band stretched between two pencils. One pencil is used as a cylindrical pulley and the sharpened end of the other as the conical pulley.

If a double coned pulley is used the belt will automatically centre itself on the maximum diameter as at this point the climbing forces balance. The self-centering action will remain so long as the cone angle exceeds the worst belt misalignment angle. With normal alignment accuracy 2 degrees half angle is ample.

It is not necessary for the two cones to meet at a sharp obtuse edge in the centre of the pulley face and the stress on the belt is reduced if the two cones are blended by a large radius. Provided the cone angle at the two sides of the pulley is maintained this makes little difference to the self-centering action and in most small pulleys the "crown" takes the form of a single convex radius extending over the whole width of the face.

For small very light duty applications circular section thermoplastic cord is useful. This is used with V groove pulleys – about 40 degrees wedge angle. The great advantage of this material is that the cord can be cut to length with a razor blade. A heated blade is then

130

placed between the two ends and as soon as the plastic has reached the temperature at which it starts to flow the blade is withdrawn and the two ends pressed together. With care a clean joint results which is almost as strong as the parent cord.

Another useful light duty belt is the polyester film belt. Polyethylene teraphalate, variously known as Mylar, Terylene or polyester drafting film, is readily available from most suppliers of drawing office equipment. This film is typically 0.003in/0.07mm thick and extremely strong and flexible. Belts can be manufactured from this by first trepanning out a large diameter "washer" with a suitably sharpened pair of dividers. This is then slipped over a circular former which is split across a diameter. Wedges are driven into the split to expand the circular former until the inner diameter of the "washer" is a few per cent greater than the original outer diameter. The whole assembly is then baked at 180 degrees C/350 degrees F to set the belt in its new shape. These make very satisfactory endless belts for small mechanisms. Their principal disadvantage is the rather low friction coefficient between the belt and metal pulleys.

All the above belt drives are friction driven and, although the drive can be tensioned so that no actual slip occurs, the distortion of the belt surface as it enters and leaves the driven members results in a small amount of "creep" which varies with load. The effect is small – typically less than 1% change in speed ratio but it means that this type of drive cannot be used if a precise relationship is necessary between the input and output shafts. Timing belts are a way of overcoming this difficulty.

These are essentially flat belts with teeth moulded on the driving surface. These connect a pair of toothed sprockets and maintain a tooth for tooth angular relationship between input and output shafts. Typical timing belts and sprockets are shown in figure 8-8.

While timing belts have major uses that need their unique zero slip capability, their high performance and general convenience in use suits them to a wide range of general power transmission applications. Some types are capable of operation at high speeds – over 8000 feet per minute/40 metres per second while transmitting substantial amounts of power. An additional advantage is that the positive tooth drive makes it possible to operate at short fixed centre distances with no provision for belt tension adjustment.

The flat toothed belt has no self-centering action and at least one of the pulleys must be provided with flanges. For most drives one flanged pulley is sufficient but if the shafts are vertical or the shaft spacing is large both pulleys should be flanged.

A point that needs watching carefully with these drives is the overload performance. The V, flat and round belt drives are all friction drives. If at some point in the operating cycle there is a temporary overload the belt will slip. Provided the overload does not persist long enough to cause overheating no great harm will be done and the drive will continue to operate normally. This is not the case with timing belt drives – these are all or nothing drives with no benign slip mode. If overloaded beyond their drive capability, permanent damage is likely to result to the belt and, in some cases, to the sprockets. Because of this, adequate safety factors must be applied to timing belt drive ratings to ensure that the overload point is never reached.

8.5 Fusing and protection

Fractional horsepower motor installations should always include suitably rated fuses or circuit breakers in the connection to the central electricity supply. However, the main function of these items is to protect the wiring in the event of a catastrophic short circuit — they will give little or no protection to the motor itself. The basic difficulty is that these items must withstand the starting current surge taken by the motor — typically two to ten times full load current. They will fail to protect the average motor which may well be damaged by a sustained overload of less than 150% of full load current.

Satisfactory protection can only be obtained if the protection device behaves in the same way as a motor, i.e. it will allow a large initial current to pass, provided it is only present for a short time, and it will interrupt a considerably smaller overload current if this is present long enough to overheat the motor windings.

Thermal circuit breakers can give this sort of protection. They consist of a few turns of resistance wire wound round a bimetal strip and connected in series with the load. Because of the different expansion rates of the two metals in the bimetal strip, it bends when it is heated up by the resistance wire and this opens contacts which break the circuit to the motor. The heater plus bimetal is arranged to heat up at roughly the same rate as a typical motor and because of this protects the motor from both short term and long term overloads.

It does not act fast enough to protect the wiring from the effects of a short circuit, so a conventional fuse or a fast-acting magnetic circuit breaker must still be included.

Thermal circuit breakers can be bought as separate items but it is usually more convenient to use a push button motor starter unit (see figure 8-9) which contains both the thermal trip and also a push button controlled contactor for switching the motor on and off line. These can be obtained for single phase or three phase operation and most contain a calibrated adjustment which enables the trip current to be set to a

Fig. 8.8 *Timing belts and sprockets*

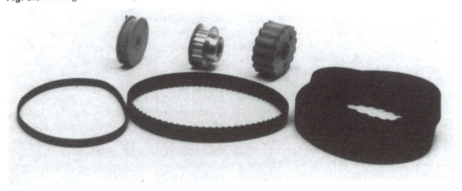

Fig. 8.9 *A typical motor starter*

value appropriate to the motor in use.

These also perform another useful function usually called "No volt release". Machine motors will stop if there is a power failure or if, for some reason, the main power distribution is switched off. With a straightforward on/off switch controlling the motor the machine will restart as soon as power is restored. In many cases a machine which unexpectedly restarts can be a safety hazard. The use of the contactor in

the motor starter avoids this problem. The motor can only be started by pressing the start button which then closes the contactor. The contactor stays closed for as long as power is present or until the stop button is pressed. If the power fails, the contactor opens. It cannot reclose when power is restored until the start button is again pressed to start the machine.

An alternative method of protection is the inclusion of thermal protective devices within the motor casing. These either take the form of a bimetal thermostat which snaps open if the motor gets too hot or a non-linear resistance element buried in the winding. The non-linear resistor, usually called a "posistor" has the unusual property of remaining at a fairly stable low value until the temperature reaches a trigger level when it increases in value by several hundred times. This is connected in series with the winding and behaves in much the same way as a thermostat with no moving parts.

These are very convenient and effective ways of protecting small motors. Some have a reset button which must be pressed before the motor will restart after recovering from an overload. However, some types will restart without warning as soon as the motor cools down and must be treated with caution.

CHAPTER 9

Identifying and Using Scrap Motors

9.1 General

Brand new motors, particularly in the larger sizes, are horrendously expensive and, for many small workshop applications, quite unnecessary. The local junkyard is a goldmine of suitable items that can be liberated from scrap industrial and domestic machinery at a price that is rarely as high as ⅒ of the cost of the new item. This chapter gives advice on separating functioning from faulty equipment and comments on the characteristics of some of the more commonly encountered items.

9.2 Picking out the good ones

Electric motors, particularly induction motors, are very reliable devices. The great majority reach the junk yard in good working order usually because the machine they are fitted to has become obsolete or has developed a major fault.

The minority that fail also reach the junk yard – this time in a pile ready to be sold for their high copper content. The junk dealer will also add to this pile motors he has removed from scrap machines. He will not have been delicate in his removal methods. Sure signs are fixing bolts and hardware cut through with an oxy-acetylene torch, mechanical damage and wires still attached to the terminals and cut through with torch or wire cutters. In contrast the dud motors will apparently be in much better condition, having been carefully unbolted from their parent equipment and having had all wires to the motor disconnected at their correct termination points.

The message is clear – be very wary of clean and tidy motors unless you can see clearly why they have been scrapped (the commonest repairable fault is dud bearings). Motors that have been forcibly extracted with little regard for minor damage are a safer bet. Better still is to remove a motor from a machine before it has been broken up, but do make sure you have the owner's permission and agree upon a price before you start attacking his hardware with a spanner.

Motors in discarded hand tools are in a different class. These rarely have proper thermal cut-outs and are frequently severely overloaded by their owners. The result of this is that a high proportion of electric drills and similar items that reach the scrap pile contain burnt out motors and are seldom worth reclaiming.

If you can find a motor nameplate, check the ratings carefully – most machines will operate from normal workshop supplies but there will be the occasional military weirdo that needs something strange like 180v 500Hz to make it function. Motors in domestic machines rarely have useful nameplate information and it is necessary to make intelligent guesses based on size and application. Make a note of the nameplate details and serial number of the complete machine – this will be useful if you later need to buy spare parts.

Most industrial machinery is fitted with three phase motors. If you are aiming to use these on single phase supplies make sure they are the six terminal type that can be reconnected in delta.

Look carefully around the machine for any motor-related items such as start or run capacitors and current sensitive starting relays. If possible remove these with the wiring still attached to the motor - it will save much head scratching later when trying to sort out the connections.

If it is a commutator motor check the commutator and brushes. Stuck or worn brushes are easily replaced but a badly worn commutator is not worth reclaiming – it's easier to wait and find a better one. Unless the motor is almost new and still shiny copper, the commutator surface should be an even blue black colour. Reject if any commutator bar shows signs of sparking or is coloured differently from its neighbours – these are sure signs of a dud armature.

Sniff around the windings (literally!).

Partly burnt insulation has a characteristic smell which is easy to recognise and hangs around inside the motor for months after the event.

9.3 Sorting out the ratings

Industrial motors mostly carry helpful name plates so identification is not a problem.

With domestic or automobile motors an obscure type number is about the best that you can hope to find. However, intelligent guesses based on size, original use and a few measurements can come close to the right ratings and disappointments will be rare.

The first useful detail is the supply voltage. Domestic machines are almost invariably rated for the local power supply voltage and frequency. Automobile items are usually 12v with 24v items common in trucks. Military items mostly operate from 24v D.C., 115v 400Hz single and three phase or 115v 60Hz.

The motor power rating is more difficult. The first step is to measure the winding resistance. With a commutator motor measure at the motor terminals. With a three phase motor measure across any pair of terminals. With a single phase motor measure the main winding only – the starting winding should either be disconnected or the starting contacts held open.

As a general guide, at full load, the current drawn by a motor in the one to five horsepower range will drop about 5% of its supply voltage in its own winding resistance. Smaller motors are generally less efficient and will drop a larger fraction – up to 10% at ¼ H.P. rising to 15% and higher at lower powers. These are *not* precise figures. They depend on the motor efficiency

135

Fig. 9.1 *3 h.p. and 1 h.p. induction motors*

and the way the losses are distributed. This can vary widely but nevertheless the method is considerably more accurate than a guess based on motor size and weight. Figure 9-1 is an illustration of the dangers of reliance on size and weight. The large motor is a one horsepower 3 phase 1425 R.P.M. machine. The smaller motor is a three horsepower 3 phase 2,800 R.P.M. machine – three times the power output of the larger machine but both smaller and lighter!

Figure 9-2 shows how motor power varies with winding resistance for a selection of common supply voltages. It uses nonlinear scales (log-log scaling) because this makes it possible to cover a wide range of motor ratings and has the side benefit that the resulting curves are almost straight lines.

The same warning on limited accuracy still applies. As a matter of interest, plotting the winding resistances of the motors in figure 9-1 on

this graph indicated output powers of 0.7 H.P. and 2.6 H.P. Sizeable errors but correctly identifying the higher power motor in spite of the misleading differences in size and construction.

If you are running near maximum rating and higher accuracy is needed then the way to do it is to measure the winding temperature with the motor fully loaded – this is covered in Chapter 10. However, this really breaks one of the cardinal rules for using motors of unknown origin – buy a large motor and run it well within its power ratings! Unless you are buying a really large motor, the size makes little difference to the price and it's better to be safe than sorry. This is particularly true if you are buying three phase motors and intend using them on single phase supplies. The larger motor will always give better starting torque.

Most of the above discussion has been about induction machines which run at modest speeds (under 3,600

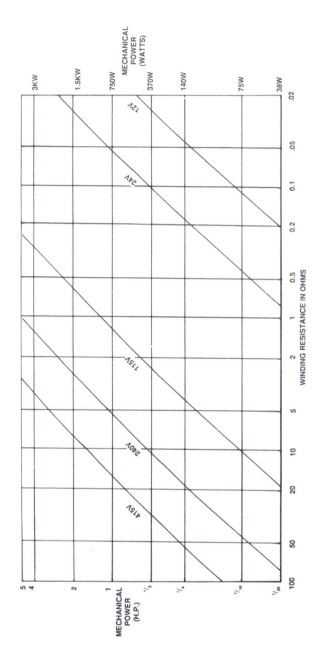

FIGURE 9.2 MOTOR POWER RATING

137

R.P.M.) with most of the loss in the copper and iron circuits. The method will give reasonable results on low and medium speed commutator motors but is not suitable for high speed machines which have a different distribution of losses.

Shunt wound commutator motors can be cross checked by a different method. These are usually designed so that, at full load, the winding losses are roughly equally divided between the armature and the field. If the rated supply voltage is known the field dissipation can be measured. Full load current will then be the value that dissipates an equal amount of power in the armature resistance.

Low voltage commutator motors will have a very low armature resistance. This can be difficult to measure and also complicated by the non-linear voltage drop across the brushes. Methods of dealing with this are covered in Chapter 10.

9.4 Common Motor Types

An enormous range of motors finishes up on the scrap pile - far too many types to be covered here. However, the two commonest sources of motors are domestic machinery and automobiles. The following comments may be useful.

9.4.1 Freezers and refrigerators

Hermetically sealed compressor units usually contain 1/10 to 1/3 h.p. 1425/1725 R.P.M. split phase induction motors of open frame construction. Starting is by an external current sensitive relay, often with bimetal thermal overload trip. Motor current consumption and/or wattage rating is often stamped on the casing.

Not very useful as a stand alone

motor but the complete assembly can be used as a vacuum pump or as a light duty compressor for air brushes. The current sensitive relay can be used as a starting relay for three phase motors converted to single phase operation.

9.4.2 Washing machines

The older types of automatic washing machines used induction motors for the main drum drive, typically 1/5 to 1/3 H.P. split phase or capacitor start 1425/1725 R.P.M. machines. Some were fitted with a multi pole second winding for low speed drive of the drum during the wash cycle. However, the power output in the low speed mode is not high enough to be a useful alternative to the main high speed mode for most applications. Normally in good condition and useful for driving a small lathe or drill press. The shaft speed isn't really high enough for a direct drive bench grinder but ideal for belt driven grinders – step up ratio 2:1 to 3:1 depending on wheel size.

Later automatic machines use a series wound commutator main motor. These are high speed motors and, although physically smaller than the earlier induction motor types, develop considerably more power. The larger frame sizes fitted to high "spin" speed machines are typically 1/2 to 1 H.P. rating. Drum drive is normally by a POLY-V belt as this is one of the few belt systems that will handle this sort of power at the very small drive pulley diameter necessitated by the higher main motor speed. Drum spin speed is about 500 to 1000 R.P.M., later machines favouring the higher end of the range. This, multiplied by the pulley size ratio, gives a good indication of the motor rated full load speed – usually about 8,000 R.P.M.

These motors often include a small

A.C. tacho-generator at the commutator end of the shaft which is used as part of the main speed control system. It produces an A.C. voltage proportional to speed but does not generate a significant amount of power. In addition a thermal sensor buried in the field winding is used by the control system to shut down the motor if it is overloaded and starts to overheat.

These motors are used with sophisticated thyristor or triac wide range speed control systems which can form the basis of an excellent lathe or drill press variable speed drive. Unfortunately there is little standardisation in the types of module used and most of the interconnecting wiring is inextricably mixed up with the main machine control wiring. Unless you are an electronics buff you are unlikely to be successful in transplanting this to the workshop environment.

However, all is not lost, as these motors will operate reasonably well from domestic power with a simple thyristor or triac power controller of the type discussed in Chapter 8. DON'T FORGET that these are series wound motors that can overspeed. They must NEVER be run at full mains voltage without sufficient load applied to the shaft.

These motors can be used for driving lathes, drill presses and small milling machines but, because of their high shaft speed, two or more stages of belt drive speed reduction will be needed. Standard V belts are not satisfactory at high speeds on small pulleys and it is better to use a POLY-V or timing belt for the first stage reduction.

Apart from the main motor there is usually a small shaded pole motor which drives a pump to empty the main drum. This is useful as a coolant pump. The shaft seals are not designed to work in contact with oily fluids but unless badly worn to start with they are usually satisfactory.

9.4.3 Cassette recorders and Hi-Fi equipment

These mostly contain small low voltage commutator motors or semiconductor-driven brushless motors. The commutator motors are designed for quiet operation and to give a long troublefree life when operated at a low power level. Because of this they mostly use precious metal wire brushes instead of the more robust carbon brushes. These wire brushes behave very well at their intended power level but will fail in a very short time if any attempt is made to operate the motor at high power levels.

Although of very limited use as motors, these machines make very handy little D.C. generators, either as a fitment to a model, or as a tacho generator to measure shaft speed.

9.4.4 Vacuum cleaners

These use very high speed series wound commutator motors run to the limit of their speed and power capability. Although many of them are rated as high as 1 H.P. they can only operate at this level because of the very high speed and the howling gale of cooling air routed straight through the motor. If taken out of this environment they will only operate successfully if drastically derated. Without cooling air 24 to 50V is as much as can be safely applied. Quite apart from this, cleaner motors lead a very hard life and discarded items are likely to be burnt out or badly worn.

9.4.5 Automobile goodies

Modern automobiles are packed with a large and increasing variety of electric and electronic gadgets many of which can be quite suitable for alternative use. Unfortunately the price breaks are not as favourable. With industrial or domestic machinery the choice facing the junk merchant is the melting pot or the occasional eccentric scavenger. Anything better than meltdown price is pure profit. Automobile equipment is different – there is a thriving market in secondhand spares at about half the new price. Although most equipment eventually reaches the melting pot, potential purchasers are assumed to need the bits as replacement spares and charged accordingly.

Another point to watch for in automobile items is intermittent rating. Items such as starter motors, window winders, seat actuators, retractable antenna motors and similar items can produce startling amounts of power in relation to their size but are only rated to do this for a few seconds at a time. Unless your requirement has a fairly similar time rating they are of very little use.

Some of the more useful continuously rated items are engine cooling fan motors, heater fan motors and wiper motors. These are mostly permanent magnet machines and, although normally rated for 10 to 15v operation can be operated well outside this rate to meet special requirements. Less than 6v to more than 24v is normally quite feasible. They can also be used as D.C. generators.

The older types of wiper motors were series wound commutator motors fitted with a tapped field to permit two speed operation. More modern types are permanent magnet machines which use a three brush system. A pair of brushes in the normal position 180 degrees apart are used for low speed operation. One of the low speed brushes and a third brush set at about 120 degrees are used for the high speed wipe. Because the brush at 120 degrees only sees a fraction of the normal motor back E.M.F. more current flows and the motor runs proportionately faster.

9.5 Motor Overhaul

The first priority is to check whether water has got into the windings or the bearings. Low voltage machines are pretty tolerant of wet windings and, provided they've not actually swimming in water, no great harm will be done. 115v and over machines are more fussy. A tortuous path through hygroscopic parts of the insulation will be quite innocuous in a dry motor. However, in a damp motor, leakage current flowing through this path will rapidly degrade the insulation and lead to very early failure. Quite the easiest way of dealing with this is to leave it for a few weeks on the top shelf in the airing cupboard over the hot water tank, but the lady of the house may take a very dim view of this enterprising use of her airing cupboard and other methods are needed. The two main aims should be to get a current of air through the motor and to get some heat into the windings. Drying out can be quite rapid if the rotor is separated from the stator and the components are left in a current of warm air from a fan heater.

The next priority is to check the bearings. Plain sleeve bearings will usually be in good condition but even if somewhat worn are not likely to fail quickly. Heavy rust on the shaft is fatal but a few light pits can be tolerated. Sintered bronze bearings mainly rely on oil held in the porous bearing structure and this should be renewed by a good soaking in light machine oil. Many motors provide an additional reservoir in the form of an oil soaked felt pad or wick in contact with the outer part of the bearing – this should be similarly soaked. Don't use grease – it gums up the pores in the wick and the bearing and is likely to do more harm than good.

Ball races behave in a different way. Treated even reasonably carefully they will give a long and troublefree life. Pitting of the inner or outer races is the normal reason for eventual failure. The pitting is the result of metal fatigue and, once started, deterioration is rapid. The onset of pitting is greatly accelerated if rust or grit gets into the bearing. If the bearings on the machine feel rough it is usually better to replace them rather than try to rescue them. However, if a spirit of optimism prevails, wash them out carefully with several changes of clean kerosene until they spin freely with no gritty feeling. If there are no obvious pits in the inner or outer race tracks smear liberally with a general purpose lithium based grease (lithium based greases usually have L or LM in the type number) and hope for the best. With rare exceptions this will have to be done with the bearings still on the motor shaft as it is usually impossible to remove them without damaging the race tracks.

Removing ball races from the rotor shaft can be difficult. There is often insufficient room for the bearing puller arms to reach the inner race and they have to be located on the outer race. Bearings that have been removed by pulling on the outer race should not be re-used as the race tracks and the balls will have been damaged by the very high forces exerted by the puller.

In really obstinate cases the last resort is to first grind through and remove the outer race. Then grind two flats on opposite sides of the inner until the metal remaining is only paper thin. A twist with the flats gripped in the vice will then break it loose.

Many motors house the bearing outers in light alloy end bells. It is much easier to remove or replace bearing outers in these end bells if the whole housing is first heated with a fan heater or hot air gun. Don't overdo it – just too hot to touch is about right.

Occasionally a motor is found with the bearing outer race loose in the end plate housing - usually at the drive end. A tell-tale sign is discoloration of the outer surface of the race. If reassembled in this state the motor will be noisy and further wear fairly rapid.

The fault is caused by poor initial fit of the bearing in its housing. If not gripped firmly the outer race flexes slightly as the balls roll past the position of maximum load and this causes the outer race to literally "walk" round the inside of its housing. This is not the same as friction dragging the outer race round and the forces involved are very much larger. Attempts to solve the problem by peening over the bearing housing or by adding a set screw to apply pressure to the outer race do not cure the root of the problem and will often fail in a matter of hours. However a permanent solution is easily effected by bonding the outer race into its housing with a suitable adhesive. For small clearances an anaerobic adhesive such as Loctite 270 is convenient – for larger gaps almost any of the room temperature curing epoxy resins will prove suitable. Both bearing outer and housing should be thoroughly degreased and covered with a very thin coat of adhesive. If lowered into position and left to cure with the rotor axis vertical the viscosity of the adhesive automatically centres the bearing in the housing. This should be the last operation when finally re-assembling the motor to ensure that, when the adhesive cures, the bearing outer is correctly aligned in relation to the rear bearing of the motor.

Be sure that the bearing is in good condition before you bond it in place because once the adhesive has cured it is difficult to separate the two components. Loctite 270 is one of the lower strength Loctite adhesives and separation should still be possible. The epoxy resin bond is likely to be permanent and separation only possible by heating to about 200°C/400°F and pulling hard. This technique also works with both the high and low strength Loctites.

On commutator machines both the commutator and the brushes may need attention. If the brushes are a reasonable length, move easily in their holders and the commutator surface is not seriously worn, leave well alone. The brushes will be nicely bedded in and both commutator and brushes are better not disturbed.

If it is necessary to remove the brushes, note their original orientation so that they can be returned to their original position facing the same way in the brush holder after cleaning. Replace if worn down to less than half the original length.

Commutators rarely fail suddenly and a considerable worn commutator can continue to behave for a long time provided the brushes aren't changed. However, if it is necessary to change the brushes the new brushes will not bed down properly on an unevenly worn surface. In this case give the commutator surface a very light skim in the lathe. Concentricity is all-important and it is best not to attempt to use a chuck but to mount the armature between centres. Use a really sharp tool with a small nose radius and plenty of top rake. Most text books will tell you to undercut the mica insulation between the commutator segments after skimming. In fact low voltage motors using copper loaded brushes are often not undercut and, apart from some increase in brush wear rate, most motors will function quite happily with flush micas.

Most spare brushes are supplied with the correct concave end radius. However, if they are square ended, or if you are adapting brushes intended for another machine, they will need to be bedded in. The easiest way is to wrap a narrow strip of fine grit silicon carbide "wet or dry" paper tightly round the commutator and work the abrasive surface back and forth a few times against the brush end. With care this can be done in a fully assembled machine with the brush in its normal holder.

Carbon brushes are a design compromise that work surprisingly well. Ideally they should be very low resistance to reduce the voltage drop between the brush pigtail and the commutator surface, but at the same time they need to be high resistance to reduce the circulating current which occurs when the brush bridges two adjacent commutator segments. In practice a compromise value of resistivity is chosen, mainly determined by the operating voltage and, to a lesser extent, by the power rating of the motor. Low voltage high power motors (automobile starter motors are one example) use low resistance carbon brushes heavily loaded with copper. Higher voltage motors use progressively higher resistivity carbon mixes. Because of this, if the correct brushes are not available for a particular motor, the substitutes should be from a motor of roughly similar voltage and power rating.

Carbon brushes can be machined to size by carbide-tipped tools or by rubbing against silicon carbide ("wet or dry") paper.

9.6 Safety

With luck, at the end of this refurbishing process you will finish up with a motor fit for your next pet project. However, in many cases the motor will be an open frame design no longer in its protective outer case and now with exposed pulleys and belts.

This is an accident waiting to happen.

Think about proper protection before you start to use the motor – if you leave this aspect until the motor is installed and working it will be one of those jobs that never gets done!

CHAPTER 10

Test Equipment

10.1 General

If your interest is in the use of standard motors used in straightforward applications then it is usually possible to just connect up and switch on – no special test equipment is necessary. However, if you are aiming to use surplus unidentified motors or to use standard motors in non-standard applications then a few items of fairly low cost test equipment can help considerably. A wide range of commercial test equipment exists for motor testing – particularly dynamometers and wattmeters. No attempt is made to cover these in this chapter as their price limits their use to professional applications.

10.2 Voltage, current and resistance

The first and most basic requirement is for one or more meters to measure voltages and currents. The most versatile type is the general purpose multimeter sold by most electronic shops. The better types can measure a wide range of A.C. and D.C. voltages and currents and are also provided with scales to measure resistance. Digital and analog multimeters are shown in figure 10-1. If limited to one meter then the analog meter is the most generally useful item. It is not as accurate as a digital instrument: $+/- 2\%$ to $+/- 5\%$ is about the best that can be expected, but this is normally close enough for routine motor measurements. The key advantage is that it is very much easier to make a quick measurement of a changing voltage or current on a pointer instrument than on a digital display. This is often needed when measuring a starting current or the current taken during temporary overload. Digital meters take a second or so to settle to their final reading and if the input quantity is changing rapidly it is difficult to interpret the display.

Most multimeters are designed for electronic work and some may not have range scales suitable for motor work. On all except the very cheapest instruments there will be adequate coverage of A.C. and D.C. voltages, small D.C. currents and ohms. However, the maximum D.C. current range may be less than 1 Amp and there may be no provision for measuring A.C. current. It is possible to extend the current ranges by external shunt resistors and rectifiers but it is a nuisance and it is much more convenient if the basic meter will cover

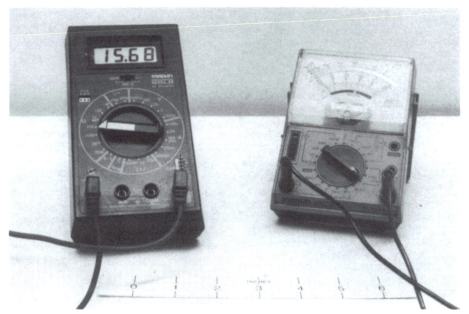

Fig. 10.1 *Digital and analogue multimeters*

10 Amps D.C. and A.C.

Meters with 10 Amp D.C. ranges are fairly common, but A.C. current ranges are only found in the more expensive semi-professional instruments. It is possible to convert the D.C. current ranges to A.C. by adding an external bridge rectifier (figure 10-2). Because the meter will then read the mean value of the current instead of the R.M.S. value (see section 1.3) it will read 10% low and the correct value of the current will be 1.11 x the indicated current. Used in this way a 10A silicon bridge rectifier is suitable for use over the current range 1mA to 10A. One disadvantage is that silicon rectifiers do not start to conduct until the forward voltage exceeds about 0.6V so that when reading an A.C. current there is a voltage drop of at least 1.2V. This is not enough to matter on a 115 or 240V circuit but can cause significant error on 6 or 12V circuits.

In many applications it is useful to be able to measure voltage and current at the same time and a second meter is a great convenience. In this case a good choice is a digital meter as the main instrument with a cheap analog multimeter for the second.

Apart from the better voltage and current measurement accuracy (+/− 0.25% to +/− 1.0%) the main advantage of the digital instrument is the greatly improved resistance measuring facility. The ohms range on an analog meter is very non-linear and, apart from a small region near centre scale, of poor accuracy. Digital meter resistance ranges are direct reading in ohms and are as accurate as the principal voltage and current ranges. This makes it possible to measure winding temperatures directly by monitoring changes in winding resistance (see 10.3).

The resistance ranges of multimeters are only suitable for measurement of resistance values down to a few tens of ohms. Few meters can be read with reasonable accuracy below ten ohms and at lower values the resistance of the leads and the variable contact resistance between the test prods and the device being measured introduce errors. If carbon brushes are part of the device then there is a further problem caused by the non-linear nature of the contact resistance between brush and commutator or slip ring – this is high and variable at the low current and low applied voltage used by multimeters.

Low resistances are better measured by the "four terminal" method shown in figure 10-3. In this a measured current is passed through the winding via the two current contacts AA and the voltage drop across the winding is measured with a separate voltmeter via the two separate voltage contacts BB. If a current of 1 Amp is chosen the meter scaling is 1 ohm per volt of reading – higher currents can be used for measurements of very low resistances. With this system the contact resistance at AA is unimportant because the current flow is measured directly by the ammeter. The contact resistance at BB is also non-critical because the resistance of the meter is thousands of times higher than any likely value of contact resistance.

10.3 Winding temperature
The outside temperature of a motor is a very unreliable guide to its internal temperature. The only unambiguous

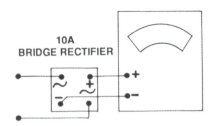

FIGURE 10.2 CONVERSION TO READ A.C. CURRENT

sign of overheating is a strong smell of burning possibly accompanied by smoke and by then it is too late! Winding temperature is the only reliable guide and is easily measured with a digital multimeter. Copper increases in resistance by 0.4% per °C. If the winding resistance is first measured at room temperature and again when the motor has reached working temperature the change in resistance is an accurate indication of the new temperature. Figure 10-4 shows the change of

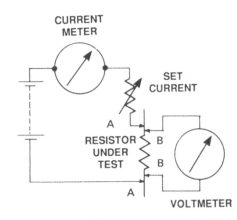

FIGURE 10.3 METHOD OF MEASURING LOW VALUES OF RESISTANCE

146

resistance versus temperature and indicates the safe temperature for different types of insulation.

10.4 Power

Power can be measured directly in D.C. circuits by measurements of voltage and current. It is more complicated in A.C. circuits because of the wattless current that flows in the motor inductance (see section 1.6).

One method of dealing with this lagging wattless current is to cancel it with an equal wattless leading current from a capacitor. This is known as power factor correction. If increasing values of capacitance are connected across the motor terminals the current taken by the

motor and capacitor combination will first decrease to a minimum value and then start to rise again. At the minimum current point, the current taken by the capacitor cancels the wattless inductive current. The remaining current multiplied by the input voltage is the true power consumption of the motor.

10.5 Speed

Speed indication is a common workshop requirement which can be met in a variety of ways. For "one off" measurements perhaps the simplest method is attach a long threaded rod to the motor shaft and note the time taken for the motor to drive a nut between two

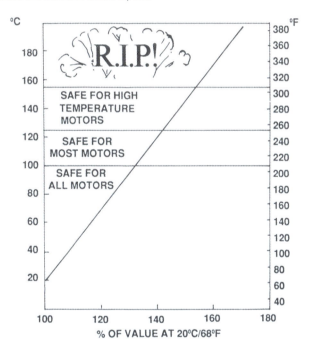

FIGURE 10.4   CHANGE OF WINDING
RESISTANCE WITH TEMPERATURE

147

marks on the rod. With suitable choice of the screw thread pitch and the distance between the marks, a wide range of shaft speeds can be measured. The method has the great advantage that shaft speed is measured directly and no special calibration is needed.

For a continuous indication of speed, small permanent magnet motors of the type used in toys or tape recorders can be pressed into service as tacho-generators. Audio tape recorder motors generate about 1V/1000 R.P.M., toy motors rather less, 0.1V to 0.5V/1000 R.P.M. depending on type.

The set up is shown in figure 10-5. Provided the generator only supplies the odd mA or so to deflect a meter, the meter calibration will be quite linear. If the resistance in series with the meter is adjusted to give the correct reading at one known speed, other speeds will then be correctly shown by the normal linear calibration marks on the meter.

Two or four pole induction motors are a convenient source of standard speed. If run with no load other than the tacho-generator their shaft speed will be somewhere between 99 and 100% of synchronous speed. Alternatively any convenient calibration speed can be measured by the threaded rod method.

Very little mechanical power is required to drive these small tacho-generators. For permanent installation on variable speed machines it is rarely worth the trouble of arranging a direct drive or gear drive. A very light duty belt drive or even a friction drive is all that is necessary. Provided it can be kept free of oil a rubber band is often sufficient. If oil contamination is unavoidable then plastic belting or an "O" ring is safer.

In some cases friction drive is more convenient. This uses a rubber-faced

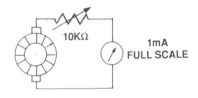

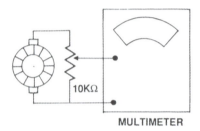

MULTIMETER

FIGURE 10.5   TACHOMETER

wheel on the tacho shaft which is then spring loaded into contact with either the side or the periphery of an existing pulley on the main machine. This is the drive arrangement used in most record players and these are a convenient source of rubber-faced wheels. If one of these is not available a quite serviceable friction wheel can be made by stretching an "O" ring into a shallow vee groove in the periphery of a metal or plastic disc.

10.6 Capacitance

A very rough check can be made on capacitors by applying a multimeter switched to its highest resistance range to the capacitor terminations. A good capacitor will give a brief deflection of the needle as the capacitor charges up and then settles down to practically zero deflection once the capacitor is fully charged. An open circuit capacitor will give no initial deflection – a short circuit capacitor will read a low or zero resistance.

This method is useful for capacitors of a few μF and larger. With smaller values the initial meter deflection may not be large enough to be noticeable.

A more accurate method is to measure the current which flows when the capacitor is connected directly across the supply. The capacitance is then given by:–

$$C = \frac{1,000,000 \times I}{2\pi FV}$$

For 240V 50Hz this reduces to:–

$$C = 13.3 \times I$$

For 115v 60Hz

$$C = 23.0 \times I$$

In each case C is in μF and I in Amps

When checking an unknown capacitor carry out the simple resistance test first to make sure that it is not a faulty short circuited capacitor that could damage the meter on the A.C. current test.

Index